ID0583543

THE LOGIC AND LEGITIMACY
OF
AMERICAN BIOETHICS

THE LOGIC AND LEGITIMACY
OF
AMERICAN BIOETHICS

Mary R. Leinhos

CAMBRIA
PRESS

AMHERST, NEW YORK

Copyright 2008 Mary R. Leinhos

This monograph was published by Mary Leinhos in her private capacity. No official support or endorsement by CDC is intended or should be inferred.

All rights reserved
Printed in the United States of America

No part of this publication may be reproduced, stored in or introduced into a retrieval system, or transmitted, in any form, or by any means (electronic, mechanical, photocopying, recording, or otherwise), without the prior permission of the publisher.

Requests for permission should be directed to:
permissions@cambriapress.com, or mailed to:
Cambria Press
20 Northpointe Parkway, Suite 188
Amherst, NY 14228

Library of Congress Cataloging-in-Publication Data

Leinhos, Mary R.
    The logic and legitimacy of American bioethics / Mary R. Leinhos.
        p. ; cm.
    Includes bibliographical references and index.
    ISBN 978-1-60497-507-9 (alk. paper)
    1. Medical ethics—United States. 2. Bioethics—United States. 3. Medical ethics—Study and teaching—United States. 4. Bioethics—Study and teaching—United States. I. Title.
    [DNLM: 1. Bioethics—United States. 2. Academies and Institutes—United States. WB 60 L5311 2008]

    R724.L415 2008
    174'.957—dc22

2008008322

To my parents, David Leinhos and Cecile Skeeles,
who taught me compassion, fairness, humility,
and the importance of singing my heart's song

TABLE OF CONTENTS

Part II:
Bioethics and the State   127

Chapter 5: Liability and Expertise:
The Emerging Professional Jurisdiction
of Bioethics in the Legal Arena   131

ACKNOWLEDGMENTS

This book would not have been possible without the help of many parties, including the National Science Foundation. I am especially indebted to the director, faculty, staff, and students at the Bioethics Center for their participation in my study, as well as their contribution of office and logistical support during my visit, and feedback on my project. I am grateful to the staff at Cambria Press for their effort in improving the presentation of my work and disseminating it. It has been a pleasure to work with you.

Heartfelt thanks go to faculty of University of Arizona, particularly Gary Rhoades for his intellectual inspiration, guidance, and warm encouragement; and to Sheila Slaughter (now at the University of Georgia) for her incisive feedback, support, and role modeling. I am incredibly grateful to Jennifer Croissant for introducing me to and grounding me in Science & Technology Studies, for bringing her wonderful track and field coaching skills to the ivory tower, and for patiently and unwaveringly supporting my intellectual development. Special thanks to friend and mentor Thomas Lindell for acquainting me with bioethics and challenging me with his good-humored and fresh philosophical perspective. To my student comrades in the Higher Education department, Cindy Archerd and Lisa Schmidt, I am incredibly grateful for seeing me through many curricular and extracurricular projects with their fellowship, friendship, encouragement, and intellectual repartee. David Guston at Arizona State University provided friendship and fostered my professional development, in addition to offering a useful theoretical framework.

In Atlanta, I thank my colleagues and friends Salaam Semaan, Marta Gwinn, and John Miles for their encouragement, and especially Hugh Potter for reviewing the pre-publication manuscript. I am also grateful to colleagues Sheila Isoke, Clyde Hart, Nick DeLaTorre, and Karen Tao for their support. From the Emerging Leaders Program, my friends and peers Ellen Olson, John Castilia, and especially Elizabeth Skillen and Krista Crider, gave inspiration and encouragement. Outside work, Ronald Chmara was a wonderful supporter of my development and success, and Christo Morehead provided sage perspective that sustained me through the writing marathon. I am very grateful for the love, humor, and support of Michael Prime during the final stages of manuscript preparation, and to Mr. P for his constant companionship for the entire duration of this project.

I have been blessed with my inspiring grandparents Ralph & Betty LaFontaine, my stepdad John Skeeles, and my parents Celia Skeeles and David Leinhos, with their unending love, support, and pride, even as three of you are sadly no longer here to witness the completion of this project. You never stopped believing in me, throughout my education process.

Thank you all, I could not have written this book without you.

THE LOGIC AND LEGITIMACY

OF

AMERICAN BIOETHICS

CHAPTER 1

INTRODUCTION

In the last 40 years, biomedical technology has not only brought humanity the ability to cheat, but indeed, to redefine death. The year 1967 witnessed the performance of the world's first heart transplantation in South Africa, which spurred reconsideration of the criteria for determining death. In due course, an ad hoc Harvard Medical School committee established a new death criterion: irreversible coma. This and other developments provoked deep concerns about the social, legal, and ethical impact of our growing biomedical capabilities. During the 1960s, several conferences, with titles such as "Great Issues of Conscience in Modern Medicine" and "Man and His Future" (Jonsen 1998a, pp. 13–19) were attended by leading figures from biology, medicine, law, theology, and the social sciences, reflecting the public's growing disenchantment with the gods of science and medicine.

Out of the growing uneasiness emerged a coordinated movement and intellectual enterprise that came to be known as bioethics, which coalesced around three new organizations: the Hastings Center (incorporated in 1969, in Garrison, NY), the Kennedy Institute of Ethics at

Georgetown University (established in 1971), and the Society for Health and Human Values (SHHV, a national professional association, founded in 1971). The SHHV's original statement of purpose suggested the goals of the new bioethics movement: "This Society sees its task as identifying these [bioethical] problems, in forming groups that will develop methods to clarify and assist in solving them, and in developing change in both professional attitudes and public awareness in relation to them" (Fox, 1985, p. 338). But what impact has bioethics made over the last 3 decades? A 1997 *Nature* article describes the growth and current prominence of bioethics, but asks "whether U.S. bioethicists have substantially shaped either the culture of science or the political decisions of recent years" (Wadman, 1997, p. 658). Bioethicist Arthur Caplan admitted, "Bioethics has a lot of authority but no real power" (Wadman, 1997, p. 658).

In this book, I examine the ways in which the legitimacy and intellectual content and organization of academic bioethics are coproduced in the United States (Jasanoff, 1996). To succeed as an academic enterprise, bioethics needs to legitimate its moral authority within the institutional structures of biomedicine and the state, and to develop the intellectual tools and content to grapple with the ethical aspects of biomedicine and biotechnology. The potency of bioethics is particularly critical today, in an era featuring a completed Human Genome Project, a thriving biotechnology industry, and daily reports of new biomedical discoveries—and new biomedical quandaries, as well as increasingly frequent litigation against clinical investigators and universities. It is hoped that this research will provide useful insight to the bioethics enterprise as it works to balance its need for institutional legitimacy with its commitment to make the biomedical enterprise accountable to all of society.

The central question of this book is the following: How are the legitimacy and logic of bioethics jointly constructed? By logic, I mean the epistemic content, approach, and social structure of the field of bioethics (its knowledge content, and the organization of the field in terms of its interdisciplinary professional work, self-described identity, and its relationships to the constituencies it purports to advise); the logic of the field is the system of expertise it provides, and the basis for that expertise.

My intent is to provide a better understanding of the nature of bioethics' impact on scientific and medical practice, and the factors that influence that impact, and, in doing so, to provide clues about how to better pursue the goals of bioethics. The investigation proceeds at three loci of analysis:

- How are the legitimacy and logic of bioethics coproduced in relation to the environmental conditions under which academic bioethics units are created and sustained? How do the conditions of institutionalization shape the approach and subject matter of emerging disciplines? This question is particularly important for fields such as bioethics, ethnic studies, women's studies, and environmental studies, which grew out of social movements in the 1960s and 1970s (Allen, 1996; Butler & Walter, 1991; Rycroft, 1991; Slaughter, 1997). Would-be disciplines must demonstrate their legitimacy among relevant constituencies, including the existing academy, to acquire the resources necessary for their establishment. Boundary work is an important part of establishing such legitimacy, and involves both staking out a new intellectual homestead and forging diplomatic relations with supporting constituencies (Abbott, 1988; Gieryn, 1999).

- How are the legitimacy and accountable expertise of bioethics jointly constructed in juristic discourse? In spite of popular interest in bioethics and the growth of bioethics programs and publications, the job market for full-time bioethicists has been limited, reflecting the immaturity of the profession. The field lacks a clear set of skills or expertise with which to identify a bona fide bioethicist (Russo, 1999; Shalit, 1997). While powerful professions can achieve formal and legal authority over their jurisdictions through state licensure, the expertise of emerging professional groups can be legitimated through other means in the legal arena. In an increasingly litigious society, professional groups are often held accountable for useful expertise via tort claims. An exploration of the liability of institutional review boards (IRBs) and health care ethics consultants provides an understanding of (1) the formal legal authority of IRB expertise, (2) opportunities for bioethics to further legitimate its research ethics expertise in the courtroom and in university research regulatory compliance systems, and (3) the ongoing construction of the accountable expertise of health care ethics consultants.

- How are the legitimacy and application of bioethics to U.S. public policy coproduced in the activities of the president's National Bioethics Advisory Commission (NBAC)? Public bioethics advisory bodies have been a staple of U.S. public policy for addressing such societal disputes, in spite of the limited direct impact these bodies have had on science and technology policy making. Kelly (2003) argued that public bioethics advisory bodies serve an important tacit function as boundary organizations that stabilize the border between science and politics, thus preserving the autonomy of science from incursion by other societal stakeholders. These boundary organizations succeed in bounding and controlling the controversy by constraining the set of issues and viewpoints that are addressed, and by dictating the decision-making strategy in ways that privilege participation of some stakeholders over others, and veil the intensity of the controversy. How do NBAC commissioners socially construct themselves and other stakeholders in the policy arena? How does boundary work performed by the commissioners and their discourse affect the legitimacy and substance of bioethics? In what ways does the NBAC itself constitute a boundary organization, establishing social order in the encompassing cultural milieu? To what extent do the democratic ideals of the original bioethics social movement remain focal?

BIOETHICS AND THE CHALLENGE OF SPEAKING TRUTH TO POWER

Bioethicists have distinguished their field from its progenitor, medical ethics. The tradition of medical ethics, focusing on the moral obligations of physicians and the appropriate doctor-patient relationship, can be traced back to Ancient Greece (Callahan, 1998; Jonsen, 1998a). The term "bioethics" was first published in 1970 (Potter, 1970), and broadened the scope of "medical" ethics to include biology generally, as well as aspects of environmental, population, and social sciences. Furthermore, bioethics is more interdisciplinary in approach than is medical ethics, drawing from law, public policy, cultural studies, religion, and the social sciences; bioethics has likewise become a fixture in the popular media (Callahan). What influence has bioethics wielded over the last 3 decades? To what extent has bioethics made biomedicine more socially accountable? At the same time, to what extent has bioethics been made into a public relations tool for academic and corporate biomedicine?

Several victories are claimed for bioethics in the protection of human research subjects and in expanding patient autonomy in medical care. Two former federal commissions with notable bioethicists among their membership are generally regarded as having been influential in these areas: the National Commission for the Protection of Human Subjects of Biomedical and Behavioral Research (National Commission), and the President's Commission for the Study of Ethical Problems in Medicine and Biomedical and Behavioral Research (President's Commission).

The work of the National Commission in the mid-1970s, in response to the Tuskegee Study scandal,[1] is credited with having transformed the climate of U.S. research, through regulations that it recommended in the Belmont Report (see Bulger, Bobby, & Fineberg, 1995). Jonsen (1998b) argued that the National Commission's work laid out the moral structure of research, making research a public enterprise.

During the early 1980s, the President's Commission issued several reports on a variety of issues, most notably "Deciding to Forego Life-Sustaining Treatment," which has been widely cited in court cases, bioethics literature, and medical ethics education (see Bulger et al., 1995; McAllen & Delgado, 1984). Furthermore, a study of state trial court judges finds that many judges with a variety of informational aids at their disposal regard ethicists' testimony as useful (Hafemeister & Robinson, 1994).

However, in spite of its past victories, steady growth,[2] press coverage and political attention, bioethics has been criticized or dismissed by many. Chambliss (1993) charged that "Traditional bioethics, at least according to the latest sociological research in medical settings, is rapidly becoming irrelevant" (p. 649). Commentators fault bioethics for disregarding the social contexts of ethical issues, decision making, and ethical theory (e.g., Bosk, 1999; Callahan, 1999; Churchill, 1999; Fox & Swazey, 1984; Light & McGee, 1998). Several observers of bioethics decry the predominant "principlism" ethical approach, which invokes four universal ethical principles (beneficence, nonmaleficence, respect for autonomy, and justice) to resolve ethical dilemmas (see Beauchamp & Childress, 1994). Principlism is faulted for being too idealistic (Chambliss, 1993), as well as "acontextual, ethnocentric, reductionistic, and sterile" (Muller, 1994, p. 451; see also Light & McGee, 1998), especially in favoring the quintessentially American principle of respect for individuals' autonomy (Wolpe, 1998).

Worse still, critics charge that the biomedical establishment has assimilated bioethics, rendering bioethics not only impotent, but even a hindrance to critical social reform. This state of affairs has been attributed to the couching of bioethical debates in medical and scientific terminology (Ettorre, 1999; Flynn, 1991a), and to the selective provision of funds to only "suitable" bioethics projects (Bosk, 1999; Evans, 1998; Stevens, 2000). Some argue that as bioethicists accepted scientific ideology (Stevens), forged a "jurisdictional alliance" with scientists over moral authority (Evans), and became professionalized, bioethics came into service legitimating biomedical progress. The biomedical establishment was thus able to defuse and redirect potent challenges from consumer and patient activists, diminishing public controversy by removing debate to an expert arena (Bosk; Kelly, 1994; Stevens; Wolpe & McGee, 2001). Furthermore, by focusing on the ethical conundrums of the affluent (e.g., managed care, assistive reproductive technology), bioethicists may reinforce interests of wealth and power, delaying progress on the pervasive social problems of the disadvantaged (Bosk; Churchill, 1999).

The activities and transactions of bioethicists are in fact coming under increasing scrutiny. In September 2000, bioethicist Arthur Caplan was named a defendant in a lawsuit alleging fraud and negligence in the recruitment of a research subject, Jesse Gelsinger, who died during a gene therapy experiment; the bioethicist had helped write the informed-consent form used in the study (Gose, 2000).[3] An August 2001 *New York Times* article raised concerns about the conflicts of interest bioethicists face when they accept donations and paid consultancies from corporations. Such predicaments are abetted by the lack of guidelines for bioethicists working with industry, and the fact that companies can pick and choose from a variety of expert bioethical viewpoints (Stolberg, 2001).

One need not search very long to find criticisms of bioethics, but locating comprehensive empirical studies that explore and verify such claims is more challenging. Although subdisciplines of sociology and anthropology are devoted to the study of medicine, social scientists have been slow to notice bioethics, with Renee Fox arguably pioneering the sociological study of bioethics (Fox, 1974, 1976; Fox & Swazey, 1974). To date, much of the sociology of bioethics has concentrated on the clinical setting, yielding ethnographic case studies of such diverse sites and issues as "medical morality" in China (Fox & Swazey, 1984), brain death and organ transplantation in the United States and Japan (Lock, 2001), the surgical ward (Bosk, 1979), and, in the United States, genetic counseling (Bosk, 1992; Ettorre 1999; Rapp 1998), intensive care units (Anspach, 1993; Zussman, 1992), alternative medicine (Frohock, 1992), and bioethics committees (Flynn 1991a, 1991b; Moreno, 1995). These studies have examined and affirmed the ways in which bioethical decision making at the bedside is shaped by its cultural, institutional, professional, and epistemic contexts.

Several scholars call for more cogent social science analysis of bioethics, and for the extension of study to other settings. Cooter (1995) found the literature on the history of medical ethics to be a "rather bloodless substitute for the political and social history of medical power, practice, and epistemology" (p. 263). DeVries (1995) lamented the dearth of attention to the institutionalization and professionalization of bioethics, and to extramedical factors encouraging the bioethics enterprise. Reminding readers that bioethics addresses not just medicine but the life sciences generally, Hanson observed that DeVries and Subedi's edited volume *Bioethics and Society* (1998) "reflects the promise and the infancy" of the sociology of bioethics (Hanson, 1999, p. 427).

Social science perspectives can help bioethicists reflect on the fit between the results they produce and their intentions. Anthropologists can refine bioethical analysis and foster critical reflection in bioethics by attending to the transformation of bioethical concepts by various constituencies, such as policy makers and journalists, and by studying the practitioners as well as the issues of bioethics (Muller, 1994). Sociology can handily counter the twin assumptions that the domain of ethics lies outside of social structure, and that "the right thinking with the right values" is sufficient to resolve ethical problems (Bosk, 1999, p. 65). Bosk provided a laundry list of questions for sociologists to ask of bioethics:

> What do bioethicists do? For whom? Under what conditions?... How are bioethicists trained? How do those in the field define their domain of responsibility? How is orthodoxy established?... How is moral authority constructed and legitimated in the case of bioethicists? How is the role and moral authority attached to it connected to an increased concern for ethics in other societal domains? (p. 66)

A small group of social scientists have heeded these calls for research, and engaged in penetrating sociological analyses of bioethics in previously unexamined settings. These studies underpin the research I describe here. In her historical study of the cultural context of the rise of bioethics, Stevens concluded that bioethics became successfully institutionalized because it diffused, rather than represented, the social challenges to biomedicine that arose in the 1960s. In its early years, bioethics shifted away from analyzing and critiquing the underlying causes of ethical quandaries, toward merely managing them; bioethics adopted "the limited role of establishing guidelines for the use of procedures and technologies that it largely accepts as inevitable" (Stevens, 2000, p. 158).

The early history of the Hastings Center, the first bioethics institute in the United States, reveals the challenges of remaining an independent voice and stalwart critic to biomedicine in the face of constraints stipulated by powerful biomedical resource providers. Stevens argued that the landmark Karen Quinlan case, which recognized the legal right to refuse life-sustaining medical treatment, served to bolster biomedical interests and strengthen bioethics' foothold in the cultural milieu. Although it is often remembered as a challenge to physicians' authority, the highly publicized Quinlan case may be better understood as a reduction of organized medicine's liability (*In re Quinlan*, 1976).[4] Altogether, Stevens' (2000) examination of bioethical issues and events of the 1970s provides a foundation for further exploration of the cultural and institutional contexts of bioethics in more recent years. Her work invites further investigation, closing with the question, "will it [bioethics] be able to free itself from the sources that help generate the dilemmas it seeks to resolve?" (p. 159).

A few studies have examined the roles of governmental bioethics commissions in public policy and the institutionalization of bioethics. Essays commissioned for the Institute of Medicine's *Society's Choices: Social and Ethical Decision Making in Biomedicine* (Bulger et al., 1995) provide comparative analyses of the National Commission and the President's Commission (Gray, 1995), national- and state-level ethics bodies (Brody 1995), and national bioethics commissions in France and the United States (Charo, 1995). Other, sociohistorical investigations, have explored the political motivations behind the Atomic Energy Commission's precursory introduction of informed consent in the 1940s (Moreno & Hurt, 1998), the President's Commission's review of human genetic engineering in the 1980s (Evans, 1998), and the role played by expertise in the workings of the Human Fetal Tissue Transplantation Research (HFTTR) Panel in the 1990s (Kelly, 1994).

Evans and Kelly examined more recent federal ethics bodies as the scene of jurisdictional conflict between biomedicine and bioethics over the management of biomedical progress. Both scholars consider the relationship between the nature of panel deliberations and resource allocation.

In Evans' view, the President's Commission's deliberations on human genetic engineering exhibited a Weberian rationalization, in which the logic of the deliberations became more systematized and less substantive (Evans, 1998). This rationalization was accompanied not only by procurement of state resources by bioethicists for reproducing the rationalized ethical system, but also by the development of a jurisdictional alliance between scientists and bioethicists, both of which advanced the interests of institutionalized biomedicine. Kelly (1994) found that the deliberations of federal ethics bodies have become less representative and more technocratic as biomedicine and bioethics engaged in an ongoing power struggle over resource allocation, with the consequences of discrediting opposing viewpoints, legitimizing the biomedical establishment, and deterring institutional change. Given "the problematic role of formal bioethical evaluation in affecting the technological imperative driving medical innovation" found by both these scholars, Kelly called for "further examination of the boundaries of medical science and bioethics" (p. 315).

Bioethics has been roundly criticized for neglecting to grasp its own social context, raising questions about the field's ability to foster more socially responsible biomedicine. Sociological scrutiny of bioethics will aid bioethicists in reflecting on their own work and its implications, and will also advance our understanding of science and technology by revealing, in more detail, the relationships among ethics, science, technology, and society. Kelly (1994) argued that bioethics has provided a moral imperative for the technological imperative of biomedicine, calling for further study of the ethical work that is done in the legitimation of science and technology more broadly.

In this book I seek to (1) enhance the findings and analysis provided by prior sociological studies of bioethics and (2) examine how bioethics balances the need for institutional legitimacy with the commitment to make the biomedical enterprise ethically accountable. What follows is an account of the coproduction of legitimacy and knowledge content in recent academic bioethics, which examines the jurisdictional contests waged by bioethicists to sustain their academic departments, establish their expertise in the judicial forum, and guide the activities and recommendations of a recent national advisory body. This work highlights the complex institutional context in which bioethicists function, and the multiple constituencies to which they are accountable as they work to persuasively inform the development and uses of biomedicine.

THEORETICAL APPROACH

Science and technology studies (STS) has made considerable progress in explicating the ways in which science and technology are socially constructed, and it can similarly improve our understanding of the social construction of bioethics. In the same way that science and technology are not value neutral, bioethics (and ethics generally) is subject to its own ideological framing, and has its own devices for obscuring its ideological underpinnings. In particular, STS offers useful conceptual tools for poststructural analysis of bioethics as socially constructed knowledge, as an emerging profession interacting with other professions, as an organizational enterprise, and as a participant in the public-policy arena.

Knowledge serves a number of important social functions independent of truth-seeking and rational logic. Ideologies, as a knowledge form, most obviously function socially "to distort, justify, or mystify group positions and interests" (McCarthy, 1996, p. 5). Knowledge functions "in generating what we know social reality to be in providing us with a sense of social unity, spurious or not; in creating and sustaining forms of domination, legitimate or illegitimate; in rendering our personal lives and relations meaningful" (p. 6). Clearly knowledge serves not only a descriptive but also a constructive function, complicating the presumed relationship between our knowledge and reality. It is beyond the scope of the present discussion to ponder the epistemological implications of the social foundations of knowledge (see, e.g., analyses by Ashmore, Edwards, & Potter, 1994; Bloor, 1994; Lynch & Woolgar, 1990; Sismondo, 1996); the present investigation is concerned primarily with which knowledges dominate and why (as opposed to which ones have veracity), and secondarily with the very real consequences of the present state of the knowledge regime. In particular, it is concerned with the question of whose knowledges are constitutive of the bioethics enterprise, and accordingly, whose interests are promulgated and legitimated when bioethics participates in the construction and shaping of public opinion and policy.

Boundary Work and Professions

Thomas Gieryn (1999) adopted mapmaking metaphors to grasp theoretically the dynamics and strategies of scientific knowledge production. Perceiving science as a cultural space, "as part of enduring cartographic classifications of cultural territories that people use to make sense out of the world," Gieryn expounded a theory of cultural mapmaking, or boundary work, referring to "the discursive attribution of selected qualities of scientists, scientific methods, and scientific claims for the purpose of drawing a rhetorical boundary between science and some less authoritative residual non-science" (pp. 4–5). As with political maps of countries, cultural maps of science reflect changing boundaries, contingent on historically situated struggles. For Gieryn, boundary struggles in science constitute credibility contests wherein cultural mapmakers vie for epistemic authority, "the legitimate power to define, describe, and explain bounded domains of reality" (p. 1). The research presented in this book examines boundary work performed on the cultural spaces of bioethics and biomedicine, and the contested boundaries between these spaces.

At the start of the 21st century, professionals, including bioethicists, are among the primary cultural producers in the United States, paid to apply their expertise in various settings (Evans, 1998). Early critical examinations in the sociology of professions culminated in the monopoly model of professions (Freidson, 1970; Larson, 1977), which regards professionalization as the quest for dominant authority, wealth, and autonomy, achieved through organizational and ideological monopolization of the market. Critical of other scholars' focus on professional autonomy and single professions in isolation, Andrew Abbott (1988) argued that professions are best understood comprehensively, as a system of professions competing for jurisdictional control over tasks. For Abbott, the hallmark of a profession is possession of an abstract knowledge system, which enables redefinition, protection, and appropriation of the problems and tasks of a profession.

Gieryn argued that Abbott's model of professions provides useful tools for the study of boundary work. Abbott's jurisdiction contests, analogous to Gieryn's credibility contests, occur in three arenas: the actual work sites of professionals, the legal arena of legislatures and courts, and the public arena of mass media and public opinion. Abbott also identified a set of contextual factors that shape jurisdictional contests and settlements, a set of rhetorical strategies for arguing jurisdictional claims, and a set of jurisdictional settlement patterns. The social-worlds aspect of boundary work theory also refines Abbott's model by examining cooperative projects across boundaries as well as jurisdictional contests, and by highlighting the fluid and heterogeneous membership of groups engaged in boundary projects (Gieryn, 1995).

The research presented here makes use of composite tools from boundary work theory and the systems model of professions, examining and comparing boundary projects in bioethics in the three arenas of the academic workplace, the courtroom, and a national public-policy body. Abbott's tools for dissecting jurisdictional contests will be used to examine how professional bioethicists, jurists, physicians, scientists, and theologians conduct boundary projects, and explore the consequences for the legitimacy and content of academic bioethics.

Organizations

Organizations also play crucial roles in boundary work, providing the necessary resources for professionals to pursue boundary projects, and, at the same time, placing contextual constraints on their efforts. Social scholars have explored how professionals' power and authority is based on the rallying of material and organizational resources (Collins, 1989; DiMaggio & Powell, 1983; Fuchs, 1994; Kay, 1993; Kevles, 1995; Slaughter, 1997). Moore (1996) observed that organizations, like rhetoric and material or conceptual boundary objects (Star & Griesemer, 1989), constitute boundaries; organizations create durable sets of rules and routines that stabilize social relations. Moore described how scientists' creation of public-interest science organizations in the 1960s stabilized the boundary between science and politics by constructing a enduring representation of science serving the public interest and obscuring the political nature of knowledge production. Guston (2000) elaborated on the concept of boundary organizations by incorporating the principal-agent perspective from political-economic theory, which underscores the roles of a boundary organization in mediating between principal and agent to ensure accountability of the latter to the former.

Hackett (2001) probed the organizational perspectives of resource dependence, new institutional theory, and technocratic organization theory for their utility in studying new organizational forms in research universities, and suggests that they provide complementary explanations. The resource dependence approach explains the strategic action of organization members in terms of changes in the organization's environmental resource base; organizational actors seek stable resources from external agents, such as a government, to sustain the organization, resulting in a struggle between internal and external actors for control of the organization (Pfeffer & Salancik, 1978; Slaughter & Leslie, 1997). While it yields testable predictions, resource dependence tells only part of the story: given empirical cases of small resource exchanges, resource dependence affords an unsatisfactory account of organizational actors' considerable efforts to please external resource providers, an account devoid of reference to the semantic and cultural implications of the exchange (Hackett).

New institutional theory remedies the semiotic deficits of the resource dependence perspective, explaining the tendency toward homogeneity in an organizational field in terms of organizations' quest for legitimacy, and hence, resources (DiMaggio & Powell, 1983). However, the theory is unclear as to the mechanisms by which coercive, isomorphic pressures operate, and it overlooks power, ideology, and the impact of the pursuit of legitimacy on the content of the organization's work and products (Hackett, 2001).

A possible response to pressures for institutional conformity is to generate new organizational forms that alleviate such pressures (Croissant, 2000). Heydebrand (1989) described the recent phenomenon of the technocratic organization, a postbureaucratic organizational form marked by informal, nonlinear authority and responsibility structures; permeable conceptual categories; and flexible structure and strategies, secured by a strong intramural cultural solidarity. Hackett (2001) argued that university-industry research relations embody technocratic organizations, but lamented that Heydebrand's concept lacks causal and developmental explanation, and is silent on the impact of this organizational form.

In contrast to Heydebrand's primarily descriptive concept, the theory of academic capitalism (Slaughter & Rhoades, 2004) endeavors to account for both the process and the implications of college and university integration into the new economy. The new economy treats knowledge as a raw material and a commodity, and impacts higher education, due particularly to its global scope, its reliance on non-Fordist manufacturing approaches, a highly educated and skilled workforce, and technology-savvy consumers. The neoliberal state fosters the new economy and academic capitalism by mobilizing resources toward production functions, and by enabling individuals as economic actors instead of emphasizing social welfare functions.

Rather than viewing higher education institutions as discrete, bounded entities, academic capitalism highlights the networks of actors that link universities to other universities, corporations, and various government agencies. Academic capitalism does not see higher education institutions as being passively corporatized or subverted by external actors, but instead describes how various actors inside universities actively employ various state resources to link their academic institutions to the new economy by creating new knowledge circuits. These activities have had a significant impact on both academic research and undergraduate, graduate, and professional education, which Slaughter and Rhoades (2004) interpreted as a shift away from a public-good knowledge/learning regime (grounded in Mertonian norms and the separation of public and private sector activities) toward an academic-capitalism knowledge/learning regime (promoting the privatization of knowledge and profit taking by institutions, faculty, and corporations). There has not been a wholesale regime change; the two knowledge/learning regimes coexist in uneasy tension in the postsecondary sector.

One manifestation of academic capitalism is the emergence of interstitial organizations, such as university technology transfer and economic development offices, which manage new activities related to the generation of external revenues for higher education institutions, helping administrators and faculty tap into opportunity structures in the new economy. University centers and institutes, a longstanding organizational form, also further the aims of academic capitalism, with many centers and institutes, encouraged by neoliberal policies, promoting partnerships with industry and state entities. A substantial number of academic bioethics programs are housed in university centers and institutes, and the research I present here examines the implications of academic capitalism for bioethics.

Organizational perspectives provide considerable insight into the structural aspects of the bioethics enterprise. These theories suggest questions and approaches for examination of the coproduction of legitimacy and knowledge in university bioethics centers and federal bioethics commissions. How do bioethics organizations constitute Gieryn's cultural boundaries? What boundary work do these organizations perform? What strategies do these organizations employ to secure legitimacy and resources? What institutional pressures do they face, and with what influence on the products of these organizations? Is bioethics employed by actors in higher education institutions to tap into new opportunity structures, and how? These questions are addressed in subsequent chapters.

Poststructuralist Policy Analysis

Kevles' (1995) historical study of physics in the United States attested that the successful discipline is one that provides useful service to governments. For 30 years, bioethics has played a regular role in public-policy formation as it pertains to science and technology, most visibly in federal advisory commissions. The boundary work perspective asks what characterizes the cultural maps bioethicists produce for the policy arena, to whom in the policy arena are such maps useful, and for what purpose. Gottweis (1998) rejected prevailing political science approaches to science and technology policy, which uncritically accept science and technology as apolitical truth-seeking activities, and neglect important relations among knowledge, language, meaning, and power. Likewise, while

> conventional schools of political science privilege either actors or structures in their accounts, poststructuralist political analysis avoids such a dichotomization by offering a language or discourse-analytical perspective that acknowledges the importance of structural phenomena and contexts for the understanding of politics without reducing actors to "outcomes of structures." (p. 12)

How can discourse analysis provide an explanation of policy formation? Such a tactic seemingly fails to account for the negotiation that created the discourse, and is not privy to closed-door negotiations that produce the final policy outcome. Poststructuralist policy analysis "looks at the texts of government not only as declarations of interest or as statements of strategy but also as material practices and as strategies to create order—as representations and interventions that actually shape politics" (Gottweis, 1998, p. 333). Likewise, Foucault asserts the profound importance of discourse itself, which is "not simply a *translation* [italics added] of struggles and of systems of domination, but that for which and by which the struggle is waged, *the very power that is at stake* [italics added]" (quoted in Larson, 1990, p. 36). In essence, language constitutes politics. Thus, bioethicists' discourse is seen as *enacting* policy, in part by legitimating a political agenda that serves the interests of bioethics, its resource providers, and jurisdictional allies. Legitimation is accomplished by a process of inscribing events and artifacts, such as embryonic stem cells, with political, moral, and social meaning.

In summary, legitimacy and knowledge in bioethics are coproduced as the result of jurisdictional boundary projects pursued by bioethicists and other professions in various societal arenas and organizational contexts.

Theoretical perspectives on boundary work in science, professions, organizations, and poststructural policy analysis provide functional approaches to revealing the nature of bioethics boundary projects and the societal interests consequently served or slighted, thus affording a means to begin assessment of the role of bioethics in cultivating socially responsible biomedical science and technology. It is my aim in this book to provide such an account of legitimacy and knowledge production in academic bioethics practiced in the United States.

Legitimacy and knowledge production, and ethics generally, are of direct but insufficiently explored interest to both the fields of higher education studies (HES) and science and technology studies (STS). There has been little HES analysis of higher education's role in shaping the larger culture through knowledge production, other than episodic philosophizing about institutional missions. In his career retrospective, Burton Clark (2000) called for more organizational analysis of disciplines and academic departments, asserted the relevance of studying science as a social institution, and admonished a fixation on an ecology-of-education framework that obscures the agency of higher education organizations. STS has generated considerable critical scholarship on the politics and sociology of scientific knowledge, but has tended to neglect ethics (for an exception, see Chubin & Hackett, 1990) and the institutional conditions of knowledge production in universities. Here I begin remedying these overlapping knowledge gaps in the fields of HES and STS by examining the coproduction of legitimacy and intellectual development in bioethics, in light of the institutional contexts of bioethics. It is hoped that this book will provide useful insight to the bioethics enterprise as it works to balance its need for institutional legitimacy with its commitment to develop effective tools for making the biomedical enterprise ethically accountable to all of society.

METHODOLOGY: OVERALL APPROACH AND JUSTIFICATION

The central question of this book is the following one: How are the legitimacy and logic of bioethics coproduced? Analysis of this question proceeds on three levels: the establishment of bioethics academic units in higher education, the construction of accountable bioethical expertise in juridical discourse, and boundary work performed in the federal public-policy discourse created by the National Bioethics Advisory Commission.

My research relies heavily on the techniques of discourse analysis and field study. Fieldwork has been used extensively in science and technology studies (STS), first in the single-site approach, as employed in laboratory studies of the manufacture of scientific knowledge (Knorr-Cetina, 1981; Latour & Woolgar, 1986); more recently multisited ethnography has been used to examine macrolevel knowledge politics, such as AIDS activism and research (Epstein, 1996) and the emergence of the trope of immunity in the United States (Martin, 1994). Fieldwork has long been valued for "thick description" (Geertz, 1973) from extensive observation, suggesting that multisited ethnography might sacrifice understanding of the tree for understanding of the forest. However, as Marcus (1998) noted, the "cultural logics so much sought after in anthropology are always multiply produced, and any ethnographic account of these logics finds that they are at least partly constituted within sites of the so-called system" (p. 81).

Thus, a multisited fieldwork approach permits analysis of how the legitimacy and content of expertise in bioethics are mutually produced at multiple sites in the American knowledge economy. This investigation "follow[s] the people," as Marcus described (1998, pp. 90–91), or more specifically, it follows bioethicists, as they seek to legitimize and expand their expert knowledge at various sites of the system—in university bioethics departments, as expert witnesses waging jurisdictional contests in the courtroom, and in the federal public-policy arena. This approach highlights the importance of social and organizational contexts in the negotiation of knowledge content and legitimacy, and reveals actors' negotiation tactics.

Discourse analysis is closely associated with poststructuralism, which perceives language as constitutive of social reality. Actors shape reality through the use of socially constructed languages, constrained by the social milieu. Scholars have used poststructuralist discourse analysis in several settings to explain how legitimation is accomplished. For example, Slaughter (1991) used discourse analysis to demonstrate how the language of official ideology of higher education, as expressed by presidents of Association of American Universities member institutions, legitimates a conservative domestic public-policy agenda in the service of private interests. Gottweis (1998) and Wright (1994) conducted post-structuralist discourse analysis of biotechnology policy in the United States and Europe. Recently, scholars have further used discourse analysis to show how legitimation strategies affect the intellectual approaches of bioethics (Evans, 1998; Stevens, 2000) and Afro-American studies (Small, 1999). Jasanoff (1998) and Halfon (1998) used discourse analysis to study the politics of expertise contained in testimony about DNA evidence during the O. J. Simpson criminal trial and other cases. In the study of legitimation and knowledge production in bioethics, discourse analysis allows deconstruction of rhetorical and ideological strategies of legitimation, and reveals the language categories shaping the character and trends of bioethics as intellectual constructions.

This study uses a multisited fieldwork approach to examine the coproduction of the legitimacy and content in bioethics knowledge, as they occur in postsecondary academic units, court documents and transcripts, and the federal public-policy arena. Data used include the primary bioethics literature; departmental Web sites, archives, and documents; interviews; professional association materials; National Science Foundation (NSF) institutional research rankings; juristic discourse; and National Bioethics Advisory Commission publications and meeting transcripts. Questions for analysis include the following: How do bioethicists portray their own profession in relation to potential resource providers, other professions, and the institution of biomedicine? What rhetorical strategies are employed to legitimate the authority of bioethics? How do such rhetorical strategies shape the approach and content of bioethics knowledge, and its ability to influence biomedicine? Which stakeholders engage in the rhetoric, and in what relationships to the developing field of bioethics expertise?

To quote Andrew Abbott's (1991) characterization of the professionalization process, the boundary work performed in the legitimation
and knowledge production of bioethics is best seen as "the multilevel,
contagious, complex social process that it actually is," not as "a simple
collective action by a cohesive group" (p. 380). Hence, the three levels
of analysis explored in this study provide interrelated, complementary
accounts of boundary work in bioethics, yielding clues about how the
legitimation and knowledge produced are mutually reinforced in the
three arenas studied. Some bioethicists have entered all three arenas, and
comparing their corresponding activities will illustrate the complex societal context of academic bioethics. The different organizational cultures
of the three arenas suggest the probability of slightly different legitimation strategies. Are these strategies consistent with each other, and with
the aims of the field? Given that both the mesolevel and macrolevel of
analysis are constructed at the microlevel, and that the microlevel units
in this study (academic bioethics departments) are the main and most
permanent affiliation of bioethicists, we are led to ask whether academic
units play a special or greater role in boundary work than the other two
arenas, (namely, the judicial arena and the federal advisory-body arena)
and hence whether academic bioethics departments warrant more attention than they receive from policy makers. This investigation's multilevel, qualitative analysis of bioethics aims to enhance our understanding
of the challenges and opportunities bioethicists face in their efforts to
shape the practice and application of biomedicine for the good of all.

ENDNOTES

1. From 1932 to 1972, the U.S. Public Health Service conducted the "Tuskegee Study of Untreated Syphilis in the Negro Male," performed on 600 African American men without their informed consent. Most were illiterate share-croppers, and about two-thirds of them had diagnosed cases of syphilis. The men were not given appropriate medical treatment for their condition, informed of the study, or given the option of leaving the study, even after penicillin became the recommended treatment in 1947. The study came to light in a front page *New York Times* story in 1972, resulting in public outcry, a federal investigation, termination of the study, and a class action, which resulted in a settlement providing monetary reparations to the sur-viving study participants, their wives, and children. See Jones (1993) for a detailed account of the study and the social conditions that enabled it.

2. PubMed medical literature database citations to ethics grew from 253 in 1965, to 1867 in 1986, and 6487 in 2005. A 2001 survey of bioethics gradu-ate training programs found that of the 65 programs that indicated what year their training program was launched, 42 were established from 1990 onwards (ASBH, 2001). As of fall 2007, the American Society for Bio-ethics and Humanities, the sole national professional bioethics association, had approximately 1500 members (ASBH, 2007).

3. Caplan had advised the gene-therapy researchers to conduct the trial on adult subjects with a mild form of the disease rather than terminally ill children, reasoning that parents of such children are unable to give truly informed consent. He was not the primary defendant in the complaint that was filed, which also named the University of Pennsylvania; the corporate sponsor of the trial; the principal investigator and founder of the corporate sponsor; two attending physicians; the CEO of the university medical sys-tem; and two children's hospitals as defendants. The case never went to trial and was settled out of court for an undisclosed sum. Caplan and the health system CEO were dropped from the suit in the settlement (Weiss and Nelson, 2001). The full civil complaint was retrieved January 21, 2008, from http://www.sskrplaw.com/links/healthcare2.html.

4. In 1975, 22-year-old Karen Ann Quinlan suffered irreversible brain dam-age from respiratory failure following consumption of alcohol and tran-quilizers in a fasting state. Karen's parents filed a lawsuit when hospital officials refused to remove Karen's ventilator. The New Jersey Supreme Court awarded Karen's father legal guardianship to make decisions about Karen's medical treatment on her behalf.

PART I

BIOETHICS IN THE UNIVERSITY COMMUNITY

How do the conditions of institutionalization and the pursuit of institutional legitimacy shape the approach and subject matter of emerging academic fields? This question is particularly important for fields such as bioethics, ethnic studies, women's studies, and environmental studies, which trace their roots to social movements in the 1960s and '70s (Allen, 1996; Butler & Walter, 1991; Rycroft, 1991; Slaughter, 1997). Would-be disciplines must establish their legitimacy among relevant constituencies, including the existing academy, in order to acquire the resources necessary for their formation and stability. Boundary work is essential to establishing such legitimacy, and involves both staking out a new intellectual homestead and crafting diplomatic relations with supporting constituencies (Abbott, 1988; Gieryn, 1999). Accordingly, boundary work shapes the logic of the emerging field.

How do institutional conditions shape the emerging academic field of bioethics? What jurisdictional claim are bioethicists making in the academy? What value do master's degree programs in bioethics provide to students, and to the academic bioethics centers that offer them? The next three chapters will address these questions.

Universities represent the primary workplace and institutional home of the bioethics jurisdiction within the system of professions. The academic wing of bioethics, like the academic wing of other professional fields, serves several functions, including knowledge production, education, and legitimation of the field at large (Abbott, 1988). Legitimation is accomplished by tying the professional work of the field to larger societal values, and communicating those links in professional and public discourse.

In order to examine the legitimacy and logic of bioethics in the academic sector, I conducted a case study of an academic bioethics center and its master's degree program in bioethics. I supplement the case study data with discourse from the corpus of bioethics literature to provide a richer account and to tie my case study findings to the broader context of bioethics in the United States. Resource dependency theory, new institutional theory, and poststructuralist boundary work theory are used to generate an account of the institutionalization of bioethics and the construction of its jurisdiction in the academy in the case of the bioethics center studied.

Chapter 2 provides an account of the establishment of a bioethics center, drawn from interviews with faculty members and documents from the bioethics center's archives. Chapter 3 examines the professional jurisdiction of academic bioethics by examining the self-perceived identity of faculty members at the bioethics center, and the relationships between the bioethics center and other key constituencies on and off campus. Chapter 4 investigates the value of master's programs in bioethics, both to bioethics centers and to their students, drawing from faculty and student commentary presented in a bioethics journal, and from student interviews conducted during my case study.

Together, these three chapters explore how the institutionalization of the center, its relationship to other constituents, and the perceptions of its faculty and students, reflect the ongoing construction of an academic and professional jurisdiction for bioethics in the existing organizational field, or arena, of professions and academic disciplines. Part I of this book thereby gives an account of jurisdictional (or credibility) contests for bioethics in Abbott's first professional arena, the professional organizational work site. These chapters provide an assessment of how successfully one bioethics center has pursued a rigorous ethical critique of biomedicine, and why it has or has not been able to do so.

CHAPTER 2

ESTABLISHMENT OF AN ACADEMIC BIOETHICS CENTER: ENTREPRENEURS OF ETHICS

In the 1990s and beyond, the number of graduate training programs in bioethics has increased by at least 42, an increase of 182% (ASBH, 2001). A significant portion of bioethics programs are housed in university centers and institutes (C&Is), most of which have an institutional home in, or are affiliated with, academic medical centers. University C&Is of all kinds have proliferated in the post-World War II era, contributing to the expansion of academic research, and have aimed at mediating between the applied knowledge demands of society and the basic research emphasis of university researchers (Geiger, 1990). C&Is' spanning of the boundary between society and universities brings both benefits and challenges. While C&Is improve the relevance of university products, the particular societal needs to which C&Is respond are driven by the funding interests of sponsors (Stahler & Tash, 1994). Hence, one ongoing concern regarding C&Is is that their programmatic focus may be too driven by "chasing dollars" (p. 545).

C&I development represents a strategy pursued by university administrators on the one hand, and by academic researchers on the other. From the perspective of administrators, C&Is are start-ups incubated by the university; successful ones will survive and contribute significantly to the funding base and research productivity of the institution (FCEPRI, 2003). C&Is, as well as interdisciplinary programs, have been cited as contributing to institutional competitiveness in national research university rankings (Geiger, 1990). From the perspective of entrepreneurial researchers, C&Is serve as an opportunity to pursue interdisciplinary collaborations (also of interest to university administrators, as a research productivity strategy), and as a means to establishing an institutional foothold for emerging academic fields, such as bioethics. An academic C&I may be promoted to the more stable and prestigious organizational form of university department if it can secure stable research and tuition funding streams, arguably signifying its societal relevance and academic credibility to the larger university community (Larson & Barnes-Moorhead, 2001). Critical to academic credibility is the compatibility of the C&I's mission with the university's goals, mission, and portfolio of academic programs; the C&I should "represent a logical initiative within the university's overall research program" (Stahler & Tash, 1994, p. 550).

Slaughter and Rhoades (2004) have described a growing tension in higher education between the traditional public-good knowledge/learning regime and the newer, expanding academic-capitalism knowledge/learning regime. The latter is associated with a policy trend at various levels promoting, among other things, the growth of C&Is, which connect academe to outside constituencies, namely, the state and industry. A related manifestation of academic capitalism is the emergence of interstitial organizations, such as university technology transfer and economic development offices, which manage new activities related to the generation of external revenues for higher education institutions, helping administrators and faculty tap into opportunity structures in the new economy.

The theory of academic capitalism (Slaughter & Rhoades, 2004) describes how universities have integrated with the global, knowledge-based new economy. The neoliberal state has worked closely with industry to build the new economy, focusing on developing individuals as economic actors, rather than emphasizing state social welfare functions. Higher education institutions operating in the academic-capitalism knowledge/learning regime have benefited from privatization and commercialization policies and from regulatory changes pursued by the neoliberal state. Academic capitalism has further been promoted by reductions in state funding for higher education, slowed growth of federal research grants and contracts, and heightened consumer sensitivity to increases in tuition, all of which have encouraged colleges and universities, both public and private, to pursue corporate revenue streams.

The new economy has posed additional challenges for academic medical centers. Rising competition from managed care (see Meyer, Genel, Altman, William, & Allen, 1998; Page, 1996) and cutbacks to Medicare and Medicaid brought on by the "Balanced Budget Act" of 1997 resulted in calamitous operating losses for many academic medical centers, which had been relying on high fees for service and federal monies to offset the costs of medical education, research, and indigent care (Beller, 2000). Academic medical centers are now also faced with competition for pharmaceutical-industry research dollars from contract research organizations (CROs) and site management organizations that promise to deliver study results more quickly than their academic counterparts. Over 30 medical colleges have set up centralized clinical trial offices modeled on the private sector, in hopes of streamlining academic research and competing successfully against CROs for industry grants (Washburn, 2005). Medical faculty members are simultaneously under pressure to see more patients, and to win more research grants and contracts, in order to increase revenues.

The aggressiveness of universities and their medical schools in pursuing research dollars is reflected in lobbying activity on behalf of higher education, and in collective efforts to curtail federal regulation in favor of institutional self-policing. The higher education sector now outspends defense contractors on lobbying efforts (Brainard, 2004). After the U.S. Department of Health and Human Services issued relatively mild draft regulations on conflicts of interest in human subjects research in 2001, national higher education associations demanded their withdrawal, and succeeded in replacing them with voluntary self-policing measures (Washburn, 2005). A substantial portion of university professors who serve as members of institutional review boards (IRBs) have consulting relationships with pharmaceutical companies (Campbell et al., 2003), and according to bioethicist George Annas, IRBs are under pressure to approve studies in order to bring in desperately needed research dollars (Washburn, 2005). In 1999, University of Toronto President Robert Prichard was exposed for lobbying top Canadian government officials to revise drug patent regulations, in order to preserve a $20 million donation that the university was negotiating with a major pharmaceutical company (Deverell, 1999).

Medical schools and their faculty members are thoroughly enmeshed with the biomedical industry. Medical schools take equity in professors' start-up biomedical technology firms, run venture capital funds, set up C&I biomedical technology incubators, and extract royalties from their faculty entrepreneurs. Faculty conduct research with grants from industry, are paid to be authors on scientific articles ghost written by industry authors, hold stock options in companies for which they consult, and relinquish protocol design and ownership of—and sometimes even full access to—clinical trial data to the sponsoring companies (Washburn, 2005).

Clearly, the new bioethics C&Is of the 1990s faced quite a challenge, establishing themselves in affiliation with resource-constrained academic medical centers. How does bioethics, caught in the tension between the public-good and academic-capitalism knowledge regimes, represent a "logical initiative" within university and medical school research programs? In what ways do bioethics C&Is reflect the interests of university administrators, and the interests of academic bioethicists? To what extent does the growth of bioethics C&Is represent attempts to tap into biotechnology-related opportunity structures? Ultimately, how well have bioethics C&Is succeeded in securing a stable funding base; maintaining a firm, self-defined mission; and providing rigorous ethical critique of biomedicine? This chapter presents a case study of the development of a bioethics C&I (referred to hereafter, for simplicity, as a bioethics center) at an American university, examining how its relevance is established to stakeholders, how it secures stable funding, and how it demonstrates academic credibility.

NATIONAL PROFILE OF BIOETHICS ACADEMIC PROGRAMS

Before presenting the results of the case study, this section outlines some broad features of the institutional landscape of academic bioethics in a profile of bioethics programs housed in U.S. colleges and universities. What characteristics, if any, apply to all or most bioethics units? How heterogeneous are they as organizations? The starting point for this profile is a national graduate-program survey conducted in 2001 by the American Society for Bioethics and Humanities (ASBH, 2001), which is supplemented here with university rank and funding data collected by the National Science Foundation.[1] Note that graduate training programs in bioethics do not correspond exactly to the set of bioethics C&Is; some bioethics C&Is do not have graduate programs, and some bioethics graduate programs are supported by other academic units, such as philosophy departments.

The American Society for Bioethics and Humanities (ASBH), the sole professional bioethics society in the United States, in cooperation with the Canadian Bioethics Society, and supported by a Greenwall Foundation grant, conducted a survey of graduate bioethics programs in North America. The ASBH Status of the Field Committee compiled a list of programs and directors, and contacted those directors, asking them to complete a Web-based survey designed by the committee. The committee's report (ASBH, 2001) analyzes the responses of 47 institutions, which offer a total of 108 bioethics graduate training programs (each degree offered was counted as a separate program, e.g., one program might offer an MA, a PhD, and one or more joint professional degrees such as JD/MA and MD/PhD).

MA programs, whether offered alone or as joint degrees, were most prevalent—with a total of 63 programs being offered, followed by 19 PhD programs, 13 fellowship programs, 11 certificate programs, and 2 "other." Bioethics graduate programs reflect a diverse array of organizational units (43% of the bioethics programs were housed in university departments, 29% in academic centers, 9% in academic divisions, 4% in academic institutes, and 4% in interdepartmental or interdisciplinary units), academic domains (33% of bioethics programs were located in a college of medicine, 20% in a college of arts & sciences, 11% in a medical center, and 9% in a graduate school), disciplinary homes (29% of bioethics programs were affiliated with the discipline of philosophy, 26% with medicine, 15% with inter/multidisciplinary affiliations, and 6% with religious studies/theology), and faculty disciplinary backgrounds (20% of faculty held appointments in philosophy, 15% in medicine, 13% in law, 12% in religious studies/theology, 10% in nursing, 6% in history, and 4% in sociology).

One of the report's appendices catalogs open-ended responses on the strengths and weaknesses of the bioethics programs, as reported by the program directors. Themes in the strengths reported include inter/multidisciplinarity, grounding in philosophical theory, and the provision of clinical experience/environment to students. Thematic concerns about program weaknesses include funding, provision of practical experience, student preparation, uncertain program future (e.g., related to recruitment, funds, marketing, insufficient faculty size, or inadequate institutional support), and contribution to students' career trajectory (e.g., a neglect in cultivating publication of student work and a lack of job placement services).

In its "Points for Discussion" section, the report highlights the expansive growth of bioethics graduate training programs since 1990 (42 programs have been established during this period, representing an increase of 182%), and conveys alarm about the lack of placement data kept by programs, particularly in light of the general overproduction of PhDs in the United States, especially those in the humanities. The presence of a "significant minority" of programs indicating that they consider their MA, certificate, or fellowship programs to be adequate preparation for full-time bioethics work, and a lack of data about what full-time bioethics jobs are available, raises concerns about programs' obligations to and responsibilities for their current and prospective students. However, as clarified in a separate article by two of the committee members, it may be the case that few bioethics students are "traditional" students; instead, most are professionals already holding positions (Aulisio & Rothenberg, 2002). The report raises the question, "will the current disciplinary diversity of the field be lost to a growing disciplinary homogeneity" (ASBH, 2001, p. 8) as more degrees and programs are identified specifically as bioethics or medical humanities?[2]

In order to examine some further institutional characteristics of bioethics graduate programs, which are clearly diverse in many respects, the pool of programs from the ASBH survey was culled and augmented. First, the certification and fellowship programs were omitted, in order to focus on actual graduate degrees in or related to bioethics, that are likely to generate graduates who will contribute to the professionalization and institutionalization of bioethics in the academy. Of the programs offering graduate degrees, some of the degrees are in traditional disciplines, such as philosophy or theology, and some are specified interdisciplinary degrees in bioethics, medical humanities, or some similar designation. The five Canadian bioethics programs in the ASBH survey were also omitted, in order to analyze the bioethics programs against data on U.S. research universities, including their receipt of federal research dollars, an indicator of institutional quality and prestige.

Uncertain as to the completeness of the ASBH survey, I employed a list of bioethics graduate programs available from http://www.bioethics.net and the American Philosophical Association's *1997 Survey of Programs in Bioethics* (APA, 1997), developed by the society's Committee on Philosophy and Medicine, to find more programs.[3] Nine additional bioethics programs offering graduate degrees were identified, and added to the population.

The unit of analysis is somewhat complicated, because some programs are jointly offered by two or more institutions (e.g., Baylor College of Medicine and Rice University share the Center for Medical Ethics and

Health Policy), and some institutions offer several discernible programs (e.g., the University of Virginia offers an MA through the Center of Biomedical Ethics, an MA and PhD in religious studies with a bioethics concentration, and an MA and PhD in philosophy with a bioethics concentration). For the purposes of examining institutional characteristics, I use the institution as the unit of analysis, sometimes considering the institutions of jointly sponsored programs together as a hybrid, as noted. Altogether there are 48 institutions, and 44 bioethics graduate degree-granting programs. One, Western Michigan University, formerly offered an MA in philosophy (applied ethics with a medical ethics concentration) but no longer does; it is included in the population of institutions as the only known expired program.

Distribution of bioethics graduate programs is skewed geographically, with most concentrated in the eastern seaboard states, and in the Midwest. With six, New York is the state with the most bioethics programs; followed by California with four; Illinois and Ohio each have three; and Wisconsin, Texas, and Michigan each have two bioethics programs. States with largely rural populations and no top-ranked research universities have few or no bioethics programs.

Bioethics programs have a strong association with elite higher education institutions. Bioethics programs are strongly associated with top-ranking institutions with regard to research funding. The Center for Measuring University Performance at the University of Florida has categorized National Science Foundation data on 616 research universities into four classes based on the amount of federal research dollars they received in the financial year of 2000: more than $20 million, $5–20 million, $1–5 million, and less than $1 million.[4] These classes are hereafter referred to respectively as tiers 1 through 4. Universities with bioethics graduate programs are overrepresented among tier-1 universities, with over 74% of bioethics programs housed in tier-1 institutions; only 26% of all federally funded research universities are categorized as tier 1. Five of the institutions with bioethics programs were not listed in the Center's data tables as receiving federal research dollars. Private universities are also somewhat overrepresented among institutions with bioethics programs. About 39% of universities receiving federal research funding are privately controlled, but that group of private institutions offers about 54% of the bioethics programs.

Unsurprisingly, bioethics programs are also strongly associated with the presence of a medical school. While approximately 20% of universities receiving federal research dollars have medical schools, 62.5% of bioethics graduate programs share a campus or an affiliation with medical schools (all four of the bioethics programs that are jointly sponsored by two institutions have a medical school at one of the institutions). Of the 13 institutions with bioethics programs but no medical schools, 6 have strong research programs in the life sciences or environmental sciences (dedicating 40% or more of their research dollars to these areas), and 4 have highly ranked philosophy departments according to the National Research Council—there is some overlap between these two groups. Eight bioethics programs remain unassociated with medical schools, strong biological research bases, or high-ranking philosophy departments.

In summary, bioethics programs are diverse in their disciplinary and organizational foundations, suggesting somewhat individualized development driven by local needs and opportunities. However, bioethics graduate programs have been more likely to arise in prestigious institutions with biomedical education and research resources, and their numbers have grown dramatically since 1990. Is the presence of bioethics graduate programs a *result of* academic capitalism, a calculated *means to* acquiring more academic capital, or both? Does bioethics primarily play a supporting role to medical and other professionals, serving educational and advisory functions, or is it emerging as a distinct field with largely self-defined objectives?

CASE STUDY OF A BIOETHICS ACADEMIC UNIT

Contextualized by the national profile of bioethics programs, this case study examines the coproduction of legitimacy and intellectual development in an academic bioethics unit. Several criteria were used for selecting potential study sites. The bioethics C&I that I studied does not so much represent an "average" bioethics unit, but rather more of a model bioethics unit, in terms of successful institutionalization. I chose a relatively secure bioethics C&I at a tier-1 institution, which would provide me with a good opportunity to examine winning strategies for establishing the legitimacy of bioethics, and the relationship between bioethics and academic capitalism. Therefore, candidate academic units for this study ideally were entrepreneurial in nature (actively, even determinedly, seeking and successfully procuring external funding), employed one or more core faculty members who are prominent in the field, and offered a graduate degree program.[5] Graduate degree programs not only confer prestige and status on their sponsoring departments, they have implications for the professionalization of bioethics, and hence have been the subject of lively debate in the bioethics discourse, as chapter 4 shall examine. A key question is whether and how managerial incentives influence a unit's developmental strategies: To what extent was the bioethics C&I that I studied being deployed as a tool by its parent institution to establish external legitimacy, to tap into new opportunity structures?

Some of my respondents asked that their individual identities, and the identity of their organization, be kept confidential. Given the small number and size of bioethics C&Is and the predominance of males in the field, all of my 28 case study respondents are anonymized as female, although both male and female faculty, staff, and students participated in the study. I refer to my study site as the Bioethics Center (or simply, the Center), and its postsecondary parent institution as Letters University. One third of the faculty, and about half the graduate student body at the Bioethics Center were female at the time of my visit in 2002. In addition to the Center director, I interviewed about 54% of the faculty (38% of these were female), including several core faculty members, education program directors, adjuncts, visiting faculty, emeritus, and joint faculty who predated the Center as tenured members of other departments on campus. Excepting the Center's director, the faculty members I interviewed were evenly split between junior and senior faculty. Notably, I was able to hold an interview with only one of the clinicians on the faculty, the director of education in the medical school (whom I have given the pseudonym "Dr. Campbell" for the sake of readability). I interviewed only about 15% of the enrolled students (nearly two-thirds of these were female), due to lack of more volunteers. I also interviewed 29% of Center staff (80% of these were female), including research and administrative personnel.

Honoring the request for confidentiality requires me to exclude several important axes of variation in my analysis and presentation, including gender, race, disciplinary affiliations, and faculty rank. Similarly, I do not identify the particular topics faculty chose to study, and omit other variables that would reveal the Center's identity. In several instances, anonymity considerations prohibit the inclusion of full references to direct quotations from my sources. In such cases full and properly formatted references are not provided.

For the most part readers will be unable even to associate quoted statements made by the same individual. I expect that bioethics scholars familiar with my study site will likely be able to guess its identity, but they should be able to associate only a few interview responses with particular persons, and these revelations are not damaging.

In spite of these limitations, I was still able to answer most of my intended research questions to my satisfaction, although it would have been possible to present an even more compelling story if it had been permissible to draw more fully from all the data I collected. I was able to address some variables indirectly by using responses from participants that compare different categories of nominal variables, such as institutional type and geographical characteristics.

I conducted a 3-week field visit to the Center at Letters University to collect my data. Sources for the case study include archival materials; Center documents and publications; the Center Web site; classroom observations; and semistructured interviews with Center faculty, staff, and graduate students,[6] with a focus on the founding director. My analysis of these data first focuses on interests and resources central to the creation of the Center, and then I examine faculty concerns about securing resources, institutionalizing the Center, and legitimizing its work. What strategies are employed to secure legitimacy and resources for the Center, what institutional pressures does it face, and with what influence on its knowledge products? How does the Center's portrayal of itself reflect the concerns of resource providers?

ESTABLISHMENT OF THE BIOETHICS CENTER

What conditions enable the establishment of bioethics centers at some colleges and universities, but not at others? What factors shape the formation of bioethics curricula, and the development of new academic bioethics centers? Slaughter (1997) observed that "most American scholars of the post-secondary curriculum continue to write about curricular formation and change as if it were *internal* [italics added] to community colleges, colleges and universities" (p. 2). Rather, she argued that "continued connections with clients, sponsors, and advocacy groups external to the university are central to the maintenance of the status, prestige, funding, popularity, and sometimes existence of specific curricula" (p. 3). With respect to bioethics programs, which have grown steadily in number since 1990 (see ASBH, 2001), Kreeger (1994) contended that media, professional, and governmental attention to bioethical issues have stimulated student and faculty interest, which, together with support from administrators and alumni, have resulted in institutional and individual commitment of resources to bioethics programs. What follows will scrutinize these and other factors more carefully, providing a more comprehensive account of the development of a bioethics center, contextualized by a number of general observations about the overall population of bioethics centers in the United States and some comparative comments made by my study respondents.

Geography

Two of my respondents at the Center, both career academics, emphasized the importance of "geography" in bioethics, meaning the physical layout of the campus and the consequent configuration and intellectual permeability among disciplinary departments within the institution. For example, one respondent noted that at Harvard (which does not have a bioethics center per se, but does have a non-degree-granting Center for

Ethics and the Professions), enclaves of scholars are physically separated by the Charles River. The second respondent made the point more emphatically, stating that the close proximity of a medical school to the rest of its university campus is "the [single] most important factor in making bioethics work." Geography is also significant to bioethics at the regional level; bioethics spread from east to west in the United States, because of the concentration of medical schools in the East, providing "incubators" where "you could learn a lot about what the [bioethical] issues were."

Institutional Characteristics

Beyond geography, a university's institutional character affects the development of bioethics programs. One respondent noted that early institutional leaders in bioethics tended to be associated with educationally innovative medical schools, such as Case Western Reserve University, whereas older, more traditional medical schools, such as Harvard, have been "late bloomers" in bioethics. She explained that this was "not due to conspiracy, but merely to a lack of personnel with a bioethics vision and/or the entrepreneurial skills to make it happen." Other institutional characteristics, particularly a university's status (whether public or private), and its related emphasis (whether on teaching or research), would reasonably be expected to have a relatively straightforward influence on its bioethics activities. For example, a private university training physician researchers might be more likely to emphasize biomedical research ethics (e.g., informed consent, genetic privacy) in its bioethics activities, while a public university with a greater commitment to training medical practitioners might be more likely to foster a focus on clinical ethics (end-of-life care, allocation of scarce medical resources). Similarly, as one respondent explained, private and public universities have different "power points;" being accountable to state legislators instead of trustees and private donors allows for comparatively less freedom in academic bioethics programs.

Students

What role do students play in curricular formation? Slaughter (1997) found inadequate the standard *demographic* explanation in higher education studies, which sees curricular change as a rational response by academic institutions and faculty to accommodate student demand; in several cases, such as for African American studies and women's studies, curricular change was instead the result of forces originating outside universities. Similarly, student demand, though present, appears to have played little role in bringing bioethics to Letters University.

The founding director of the Center cited students as one of the key factors in establishing the Center, recalling that "students were agitating in medical school classes to get medical ethics into their curriculum." When probed further, she explained that by 1994, while most medical schools had medical ethics programs, Letters did not, stating "I think they [the students] just felt like they were getting left out."

However, another Center faculty member, whose faculty tenure at the university stretches back to before the director was brought in to head the new Center, dismissed the causal role of medical students. She explained that in the 1980s, a group of senior medical students developed a bioethics component for the second-year medical curriculum (consisting of four 1-hour seminars led by a medical student leader, often accompanied by faculty leaders, according to a subsequent bioethics Task Force report). Students also put pressure on the medical school's academic dean to provide more ethics training, she said, but the dean "cooled them out" with a lecture series and some underwriting for their student journal. My cynical faculty respondent observed that the demands of a transient student population are easily ignored, and that the medical school was taking credit for doing wonderful things in ethics long before the Center was established, and even at the height of student complaints.

Indeed, student demand was cited neither in the bioethics Task Force report, which called for an expansion of ethics training, nor in the subsequent proposal for a regent/trustee professorship in bioethics. When asked about the origins of the Center, the director of medical education spoke of the medical school's need to "catch up" with other universities, all but one or two of which already had medical ethics programs, or at least a half-time biomedical ethicist on the faculty. She did not mention student demand. Most likely, the primary reason that Letters medical school needed to catch up with peer institutions was the recent introduction of new accreditation requirements for medical ethics education, invoked respectively by the Liaison Committee on Medical Education in 1989 and the Accreditation Council for Graduate Medical Education in 1990. Taken together, these accounts imply that student demand was insufficient to bring about a medical ethics education program, let alone the Center, to Letters University. Accreditation requirements and institutional isomorphism were more powerful influences, capable of instigating curricular change.

It is not my contention that students have no impact on curriculum and educational practice, but rather that their influence is likely to be more incremental. For example, a faculty member's article in the Center newsletter cites two instances where student concerns resulted in changes in practice. First, medical students in the gross anatomy course objected to the course's use of unclaimed bodies from the local Medical Examiner's Office; according to the newsletter, "Feedback of these concerns to those responsible for that decision was well-received and the decision to use such bodies was reversed." In the second instance, student concerns about the adequacy of informed-consent language for a clinical trial at Letters, discussed in a pharmacology course, "refocused the attention of those responsible for the form, and helped redirect our efforts in informed consent processes." Student demand from a captive, transient audience, without accompanying consequential institutional pressure from other universities, accrediting bodies, or the state, is unlikely to impel faculty and administrators to locate and dedicate the necessary resources for developing new courses, programs, and academic units.

The Director of Medical Education

According to Slaughter (1997), the *learned disciplines'* account of curricular development in higher education studies claims that faculty members, not students, are key instigators of curricular change, where faculty as professional actors mold the progressive, rational development of scholarly disciplines. "Great Man" narratives, in which individual persons are credited with a leadership role in curricular change, are consistent with this viewpoint. However, the learned disciplines' perspective overlooks the politics of knowledge, which constructs the legitimacy of curricula in a complex organizational context (Slaughter, 1997). In the case of the Center, several respondents cited the director of medical education, Dr. Campbell, as a valiant and even crucial figure in the development of bioethics at Letters University. The Center director not only named Dr. Campbell as one of three contributory parties in establishing the Center, but also pointed to Dr. Campbell's counterpart at another university in the region, explaining, "clinicians that really grapple with ethical issues [due to their specialties] and earn lots of income for the [university] hospital can have lots of impact." Two other veteran faculty members were complimentary about Dr. Campbell's efforts in medical ethics education, but expressed that she lacked the power to bring about the creation of the Center. While they described Dr. Campbell as "heroic," "valiant," and a "major shaper of the curriculum," who had "pushed hard for a bioethics presence" and "kept bioethics teaching alive," they viewed the dean of the medical school as ultimately responsible for the creation of the Center.

While, in the end, it appears that the Center owes its existence mainly to the medical school dean and the director of the medical school's Biotechnology Institute (whose roles I will address shortly), Dr. Campbell's role in laying the groundwork for the Center and focusing persistent attention on bioethics at the medical school was indispensable. At the least, by exercising leadership in proposals and justifications for the Center, Dr. Campbell helped the Center to be created sooner than it might otherwise have been.

As a first step toward bringing the ethics portion of the medical curriculum up to speed with that of peer institutions, Dr. Campbell said she had proposed that the medical school dean create a Task Force to develop a comprehensive bioethics education proposal. A bioethics Task Force was convened with Dr. Campbell as its chair, and given the charge "to evaluate the current education in bioethics in the [medical school] and develop a more comprehensive bioethics program for undergraduate, graduate, and continuing medical education" (Task Force report).

The Medical School Bioethics Task Force

Finding that then-current bioethics offerings were "inadequate," with a "lack of overall coordination and breadth of educational content," the Task Force proposed that an academic bioethics program be developed in three phases. The program would begin with the undergraduate medical curriculum, progress to the graduate and continuing medical curricula, and culminate in the development of a multidisciplinary center in bioethics to organize and facilitate cross-campus academic collaboration. The Task Force's report recommended that the new program be integrated into existing medical courses and include both knowledge and competencies in bioethics. Furthermore, the Task Force called for the creation of two full-time bioethicist positions with support staff, so that the bioethics program would be "directed and implemented by faculty with expertise and training in the subject."

The Task Force's recommendations were supported by a "Statement of Beliefs," which justified the legitimacy of a bioethics program (and in so doing revealed expected challenges to the legitimacy of bioethics), and shed light on the Task Force's conception of bioethics. Firstly, the "Statement of Beliefs" asserted the centrality of ethics to medicine, stating that "competence in identifying ethical issues and in making ethically informed decisions *is as necessary to a physician as competence in the medical sciences* [italics added]" (Task Force report). Secondly, the "Statement of Beliefs" emphasized the authoritative *teachability* and *testability* of bioethical knowledge, affirming that ethics constitutes "*objective* topics with *educational content that can be taught and learned* [italics added]," and that the standards in bioethics education should be on par with the medical curriculum, "including examinations, minimum acceptable competencies, and grading" (Task Force report). Finally, regarding the scope of bioethics, the "Statement of Beliefs" pronounced that the bioethics curriculum should be comprehensive, including both ethical theory and its application to medical practice, and encompassing "issues related to the interface of bioethics, law, and medical practice" (Task Force report). Taken together, these stated beliefs establish the legitimacy of biomedical ethics as an academic subject, by establishing its status as objective knowledge and its relevance to and inseparability from the field of medicine. It is not clear to what extent the justifications given in the task force's report were persuasive. However, the Task Force's recommendations for a bioethics curriculum were immediately approved by the medical school and its faculty, and Dr. Campbell assured me there was widespread support for and interest in the recommendations within the medical school.

Proposal for a Regent/Trustee Professorship in Bioethics
A subsequent proposal submitted to the medical school dean, again spearheaded by both Dr. Campbell and the vice dean of medical education, sought the creation of a professorship in bioethics. The person hired to fill the professorship would play a leadership role in planning and implementing the medical school's bioethics curriculum, and in developing a new bioethics center, then under consideration. The professorship proposal expanded arguments for the legitimacy of academic bioethics beyond local curriculum and medical practice, shifting the focus toward the regulatory and funding environment of biomedicine. The proposal appealed to new accreditation requirements for undergraduate and graduate medical education in bioethics, which were invoked respectively by the Liaison Committee on Medical Education in 1989 and the Accreditation Council for Graduate Medical Education in 1990.[7] Furthermore, grant-making bodies, chiefly the National Institutes of Health, were beginning to make funding conditional upon the recipient's provision of formal ethics training programs.[8] The proposal deduced that the medical school had "the urgent need" to develop a strong bioethics program in order to compete with leading medical education institutions nationally. Citing the medical center's short-term plan, the proposal asserted the "necessity for joining new strength in bioethics" to the medical center's growing strength in molecular biology. This necessity, the

proposal noted, had already been recognized by federal funding agencies, exemplified by the 90 to 150 million dollars (these figures were underlined in the proposal) allocated for the ethical, legal, and social implications of the Human Genome Project. The medical school needed to pursue bioethics not merely out of scholarly and practical interest, but in response to environmental expectations and new opportunity structures. While Dr. Campbell and her colleagues on the Task Force were undoubtedly motivated by a scholarly commitment to doing the right thing by bringing bioethics into medical education, clearly they also recognized the utility (and likely, the necessity) of citing institutional competitiveness in the academic capitalism knowledge/learning regime in order to bring bioethics to Letters.

The Medical School Dean

Key faculty respondents provided somewhat divergent accounts of the medical school dean's motivations and enthusiasm for creating a bioethics professorship, which eventually became a reality. Delay in creating and filling the professorship was variously attributed to lack of enthusiasm for internal candidates (including Dr. Campbell) and difficulty identifying and recruiting distinguished external candidates, in part due to limited funding availability.

Based on respondents' reports, the emergent account of the dean's motivation suggests that she had little interest in bioethics, at least until funding and a compelling vision of bioethics' contribution to the medical school's interests materialized. Dr. Campbell recalled that the dean had been receptive to the professorship proposal. However, in a cost-reduction environment, the search committee was charged with getting "the best bioethicist we can afford." Another long-time faculty member stated that although in retrospect the dean is widely viewed as having been enthusiastic about bioethics, during the candidate search, which lasted more than 2 years, the dean was in fact "lukewarm" about bioethics, and did not offer much money to "recruit a distinguished person."

The search was revitalized when the director of the Biotechnology Institute offered a substantial amount of money to help establish a formal bioethics presence. According to faculty respondents, the institute director felt it was important to convey the medical center's seriousness about ethical concerns, particularly related to in-house biomedical research. The medical school dean was responsive, agreeing to cover the remaining expense, and as Dr. Campbell put it, "betting big" on biomedical technology with the Institute director. The director of the Bioethics Center explained that the medical school dean "decided to make a big commitment in the area of genetics and thought that ethics would be helpful to the growth of the genetics initiative—both for deflecting criticism, [and] for raising issues...it would be both a shield and a sword." Another faculty respondent commented that the dean "got what she wanted" out of the deal, namely a highly visible new specialty center that brought public attention, and presumably business, to the university medical center.

Academic medical centers faced tremendous environmental resource challenges in the 1990s, including a highly competitive health care market and shrinking Medicare payouts. Like many of its peers, the Letters university medical center first attempted to gain a competitive edge through expansion, acquiring hospitals and private practices to bring more patients, and thus revenues, to the system. The university medical center also cut hospital expenses by 10s of millions. The medical school dean also adopted an expansionist approach within the medical school, creating eight new biomedical C&Is (including the Center), and recruiting more than a dozen departmental chairpersons to head the C&Is and fill other positions, all during the first 6 years of the dean's tenure (Letters campus newspaper article, 1995). The dean took an aggressive approach in expanding the medical school's activities, and extended them to include bioethics once she recognized its potential to enhance the medical academic portfolio.

DEVELOPMENT OF THE CENTER:
FINANCIAL AND INTELLECTUAL INDEPENDENCE

With adequate funding provided, the bioethics professorship was soon filled, and the new bioethics center was under way. Dr. Campbell, credited with helping the new director "establish credibility" in the medical school community, also encouraged the director to seek departmental status early on. However, departmentalizing bioethics is a formidable challenge. Canadian bioethicist Christine Harrison (2002) explained, "existing university departments, chairs, and promotion committees often struggle to understand how bioethics can 'fit' into existing models. This may hamper career advancement and other academic rewards [for academic bioethicists]." (p. 20). Harrison's comments are applicable to academic bioethics in the United States as well as Canada.

Letters University's medical school requires its department-based faculty to generate overhead through grants, as do most academic research institutions. In the medical school paradigm, promotion awards are based on winning grants and publishing articles, but bioethics faculty are more likely to publish books than prestigious journal articles, and they receive fewer and more modest grants than their biomedical colleagues. Accordingly, Harrison (2002) explained, "in Canada hospitals and universities are currently unwilling or unable to make a long-term commitment to bioethics programs and services, due in part to the lack of stable funding" (p. 20). Again, the case is much the same in the United States. Would-be bioethics departments must make a compelling case for such status change and negotiate for different promotion and tenure criteria, or rely on hybrid faculty with joint clinical and bioethics appointments who can meet more traditional medical scientific standards.

Apart from the medical school at Letters University, the College of Arts and Sciences (CAS) was also unlikely to house a bioethics department. One faculty member commented that CAS probably would not take in a bioethics department because it was not a discipline; other interdisciplinary programs which had been created under special conditions, and were vulnerable, had a history of being phased out. Another respondent noted that medical school departments, on the other hand, are not necessarily disciplinary, but may be "practical" (e.g., emergency medicine). Thus, the questionable status of bioethics as a discipline per se would not hinder the establishment of a bioethics department in a medical school. Altogether, however, the financial logic of medical schools and the disciplinary orientation of colleges of arts and sciences make it difficult for academic bioethics centers to departmentalize, as demonstrated by the small number and size of bioethics departments, relative to the population of bioethics C&Is in the United States.

Rather than deal with departmentalization challenges, the new Center director pursued other strategies to foster the Center's financial stability and intellectual independence. The director recounted to me that she warned the medical school dean when she was hired that she would not be granting "protection" to the medical school. In fact, she predicted that she would more likely be a "source of aggravation," which in retrospect she felt was the case. Dr. Campbell described the Center as "organizationally under the [medical school] but functionally between schools" at Letters University, which established some "strategic distance" from the medical school and promoted intellectual collaboration with other campus organizations. The director explained that her strategy in setting up the center included acquiring "intellectual and financial capital," and some "political scoping out," to pursue the activities of education, research, and outreach. As an outreach activity, hosting conferences not only serves to address pertinent bioethics issues, it also promotes the Center's visibility outside the medical center.

Many bioethics programs receive funding from medical school overhead, and the Center is no exception. However, without departmental status, most bioethics programs and centers lack guaranteed support. The Center director maintained that the "safest" way to fund a bioethics center was to seek diversified income from four categories of source: the university; federal granting agencies; foundation grants; and private gifts from corporations, patient groups, and wealthy patrons. Furthermore, she stated that a center should not be "over-reliant on any one source," lest it be "vulnerable to pressures." She cited the need to "stay in touch with all constituencies," and the importance of openness, noting the great illusion among academics, presuming the "goodness" of government and foundation sources compared to corporate funding sources.

In spite of the Center director's diversified funding-source strategy, the freedom of faculty to study bioethical problems and issues they found compelling was constrained by the interests of funding providers from all sectors, and faculty members at the Center were openly aware of the ways in which foundation- and government-funding opportunities shaped their work. Respondents noted that their interests in producing scholarship on reproductive rights and distributive justice in health care, for example, were not attractive to foundation or government sponsors, which focus instead on genetics and other high-profile issues. One faculty member noted that extramural collaboration with scholars at other bioethics centers is attractive to funders, and also benefits bioethicists through the mutual enhancement of the collaborators' reputations. Another respondent predicted that research ethics will create an enormous demand for bioethicists, and accordingly, research on research ethics will become a primary activity in the field.[9]

Although the fledgling center was unable to pursue much independent research, a number of other programs at the medical center approached the Center about serving as a coinvestigator on various projects. While a number of collaborations resulted, providing the new Center with several organizational ties, these associations failed to secure any funds for research at the Center. According to the Center's first annual report, funding agencies valued the Center's involvement in projects, and were even "positively influenced toward making awards to [university] programs," but "the Center has been dropped from funding in every case on the grounds that ethics is or should be included in grant overhead." In response, the annual report promised the Center would change its focus to independent grants sought by Center faculty, and fundraising with private and individual donors.[10] However, a new problem accompanied the pursuit of independent grants. One of the Center's later annual reports describes the hardship created by the facilities and administration (F&A) fees imposed on sponsored research, which the university charged at a rate that was "prohibitive for many foundations that traditionally fund bioethics research." The report acknowledges Letters University for allowing proposals to go forward with less than the full F&A rate, urging that such concessions continue.

The Center's home in the medical school defines not only funding opportunities, but also the conditions of promotion and tenure of Center faculty. Faculty at the medical school were evaluated on the basis of regularly publishing articles in top-tier medical journals and securing RO1 grants from NIH. Publishing books counts for little in the promotion and tenure scheme of academic medicine; in the rapidly developing field of medicine, book contents are likely to be obsolete by the time the volume reaches library shelves. Several of my faculty respondents expressed anxiety about promotion and tenure requirements for the Center's junior core faculty, as they anticipated the eventual departmentalization of the Center. These young professionals were trained in humanities or social science disciplines that expect book writing, and not necessarily a steady stream of research grants. From the base of a formal department, bioethics faculty would gain the opportunity to directly influence committees that determine promotion and tenure criteria, but would have to communicate a compelling justification for establishing rules more appropriate to their work products. These circumstances would appear to favor the promotion of academic physicians over humanities or social science scholars among faculty, if the Center were to departmentalize within the medical school.[11]

Several faculty respondents discussed the dilemma of accepting corporate funding for bioethics. One noted that unlike the Letters Cancer Center, the Center was always questioning whether to accept corporate dollars, but while she was wary of corporate dollars, she felt it was necessary to do so. Respondents noted ways in which they actively limited any perceived or real undue influence from corporate sponsors, including limiting the amount accepted, and balancing it with grant monies from nonprofit foundations and government agencies. They also cited various measures to discourage conflict of interest, including accepting funding only on the conditions that faculty members fully disclose their industry involvement and sponsored work; retain the academic freedom to publish their research findings' and clearly express that acceptance of funding support does not entail acceptance of the sponsor's policies, products, or services.

The Center's sixth annual report acknowledged as one of the Center's major challenges the need to

> make certain that the Center uses its expertise in cutting-edge bio-ethics to first, chart the course of research, and then to find dollars to support it, rather than first finding dollars and then creating a research agenda to fit the dollars. While the Center is exquisitely suited to explore emerging areas, it is difficult to fund such areas.

The report outlines a 3-year strategic plan to put the horse in front of the cart. Notably, the plan called for the establishment of standing research programs in "fields of strength at the Center," and for the development of grant support for these long-term projects. The plan also called for the creation of "a standardized approach to respond to requests for sponsored research by for-profit organizations consistent with University protocols and academic freedom," and specified a target of five projects sponsored by corporations.

Within 18 months, the Center had made considerable progress in diversifying its funding streams. At the beginning of the period, the Center, according to the sixth annual report, was operating with 2 federal agency grants and 7 foundation grants, with half the total monies coming from the government grants. Another 18 months later, according to the subsequent annual report, the Center had secured 13 additional grants, including 3 from corporate sponsors, representing 19% of new grant revenues. The Center was awarded 5 grants from government sources and 5 from foundation sources, with government providing 68% of new grant revenues. The 13 new grants more than quadrupled the proceeds of the 9 previously obtained grants. The gift funds received by the Center at the end of the period were also diversified, with 22% donated by individuals, 33% by corporations, and 45% by nonprofit organizations; gift revenues totaled less than 7% of sponsored research revenues.

Even with a diversified funding portfolio, as mentioned previously, bioethics scholars are subject to the research and program agendas of their patrons, irrespective of which sector these come from. However, bioethics is especially vulnerable to the power of the biomedical corporate sector, which has penetrated the academy in ways foundations and government do not, at least in the United States. Bioethicist Carl Elliott (2001) explained in *The American Prospect*,

> Corporate money is so crucial to the way that university medical centers are funded today that no threats or offers need actually be made in order for a company to exert its influence. The mere presence of corporate money is enough. (p. 17)

Elliott (2001), a firm opponent of bioethicists' acceptance of corporate funds, argued that the public credibility of bioethicists "rests on the perception that they have no financial interest in the objects of their scrutiny" (p. 20). However, his observation about corporate funding of universities renders it all but impossible for bioethicists to claim no financial interest in corporate biomedicine, because the university medical centers that house most academic bioethics centers rely on corporate dollars, even if the bioethics center itself does not itself have direct involvement with corporations; there is no escape from corporate ties for these centers.

It is both essential, and a tremendous organizational challenge, for academic bioethics centers to establish a measure of organizational and financial independence, in order to maintain an independent voice and credibility. The Center director tested this independence at Letters University early in her residence at the Center. She had publicly criticized a local health care company that had made substantial financial contributions to the medical school. Consequently, the company CEO demanded the Center director's termination from the medical school dean. The dean refused, citing academic freedom and the Center's autonomy, and denying any responsibility for the Center's activities. Such deniability is important to the director, who strives to ensure that the Center's faculty feel they can speak out freely—she views it as her job to "cover them." She contended, that in general, the Center has "lots of autonomy and independence," and that neither the medical school dean nor the university president keeps track of the Center's activities. "In some ways it's better that the university president doesn't exactly know what's coming out of here [the Center], it can be very provocative or upset people," observed the director, stating that "there is a certain kind of power in being distant administratively." While this instance of preserved academic freedom is heartening, it does not assure carte blanche for the Center or its director, or that bioethicists at other universities could be as openly critical without imperiling themselves professionally.[12]

THE ORGANIZATION AND INSTITUTIONALIZATION OF THE BIOETHICS CENTER AT LETTERS UNIVERSITY

When asked about challenges to the influence and expertise of the Center, the graduate program director stated that the biggest influence on the Center was "larger institutional forces." The Center director affirmed that the biggest challenge for bioethics is its traditional home in the academic medical center: an "institution in flux," facing fiscal challenges and the managed care paradigm. Bioethics, often viewed as an overhead activity, is "vulnerable to elimination." This weakness is made more acute, she said, by the fact that bioethics "hasn't figured out professionalization."

By way of summarizing my findings thus far, I shall now discuss the establishment of the Center using the complementary perspectives of resource dependency and new institutional theory. These rather structural interpretations will be balanced with a more agent-oriented account of the identity of bioethicists and the boundaries of their jurisdiction in chapter 3.

Resource Dependency

It is clear from the preceding account that resource dependence (Pfeffer & Salancik, 1978; Slaughter & Leslie, 1997) was a key challenge for the establishment of the Center. The endowed chair necessary to spawn the Center was not created until the director of the Biotechnology Institute partially underwrote the position, which also likely guaranteed that she would have congenial relations with the new Center director, and possibly veto power in the selection of that director. Once established, the new Center's efforts to bring in grant revenues were constrained by high indirect costs charged by Letters, and by the funding conditions set forth by grantors. Center faculty's initial selection of scholarly projects was undoubtedly shaped by the collaborations they were invited to undertake with established medical school faculty in traditional biomedical departments.

The Center director was acutely aware of the fiscal constraints imposed by funding sources, and took the strategic approach of pursing as diversified a funding portfolio for the Center as possible. Other Center faculty members were also aware of the influence of funding sources on the work they selected, and all took precautions against conflict of interest in the acceptance of funding from corporations. However, as mentioned previously, corporate influence is unavoidable, due to the strong ties that medical centers have with the biomedical industry.

Institutional Isomorphism

The concept of institutional isomorphism, the tendency of organizations in the same line of business to become more similar to one another as they attempt to respond to the same set of environmental conditions (DiMaggio & Powell, 1983), provides a compelling account of why and how bioethics has, by necessity, a close and amicable relationship with professional medicine. This organizational intimacy with the medical field leaves bioethics vulnerable to charges of being a lapdog, and the institutional isomorphism perspective helps us understand why establishing and maintaining professional autonomy is a significant challenge for bioethics. Essentially, because bioethics C&Is are most commonly created and housed in academic medical centers, they resemble medical departments or other biomedical C&Is, and the work of academic bioethicists is structured in a way similar to that of their physician colleagues. While, at first glance, this observation is not profound, recognition of the depth and nature of the similarities and their implications for professional work is. The implications for academic bioethics of the three different types of institutional isomorphism (coercive, mimetic, and normative) will now be examined.

Coercive Isomorphism

The dependency of the Center on the university medical center is a tremendous source of coercive isomorphism, a type of isomorphism resulting from political pressures exerted by other organizations and cultural expectations, underscoring the problem of legitimacy (DiMaggio & Powell, 1983). The medical school's promotion and tenure requirements and overhead charges to grants generate considerable pressure on the Center to conduct its academic work as medical departments do, even though the nature of bioethics scholarship is quite different from academic medical research. However, the similarity of bioethics to traditional medical school departments in terms of sharing a practical rather than disciplinary orientation, as well as the more favorable funding environment of medical schools, makes the university medical center a more likely home for the Center than the College of Arts and Sciences. Coercive isomorphism also works to cultivate a need for bioethics units at medical centers, as new training requirements from NIH and accrediting bodies, and new funding opportunities, create a role for bioethics in medical centers.

Mimetic Isomorphism
Uncertainty about goals, organizational technologies, or the environment fosters mimetic isomorphism, where institutions model themselves on peer organizations (DiMaggio & Powell, 1983). Indeed, the need to be competitive with other medical schools was cited as a factor contributing to the establishment of the Center. One Center faculty member indicated that in controversial or complex matters, such as developing rules for working with corporations, the Center looks at what other bioethics C&Is around the country are doing. It is also worth noting that the Center director was recruited from a position at the bioethics center of another university, and her prior experiences at that center undoubtedly shaped her development of the Center at Letters.

Normative Isomorphism
The struggle to establish and control a professional jurisdiction results in normative isomorphism, such that members of a professional group are rendered very similar to one another through the socialization and standardization processes that continue throughout professional training

and practice. The medical school bioethics Task Force report, by way of justifying the inclusion of ethics in the medical curriculum, emphasized the ways in which ethics knowledge is similar to other components of the medical curriculum, in terms of objectivity, teachability, learnability, and testability; these expectations undoubtedly shaped the way in which bioethics was integrated into the medical curriculum. The 1998 consolidation of three bioethics societies into one[13] (the American Society for Bioethics and Humanities), was likely induced by, and further contributes to, normative isomorphism in bioethics. However, some factors in the bioethics organizational field may serve to counter normative isomorphism. Firstly, the interdisciplinary nature of bioethics means that bioethics C&Is will be staffed by faculty with degrees in different disciplines, and will subsequently bring an array of somewhat divergent professional norms to the table. Similarly, because bioethics is a young and therefore small field, many bioethicists are isolated, as one of my faculty respondents noted. Many bioethicists will be one of a kind at their universities, and will call a disciplinary department home, rather than a bioethics C&I. As bioethics continues to professionalize, we can expect to see more normative isomorphism, particularly as more bioethics PhD programs emerge. The professional identity of bioethicists will be examined in chapter 3, and the role of training programs in normative isomorphism will be revisited in chapter 4.

It is important to consider the ways in which being housed in academic medical centers can serve the traditional goals of bioethics, and not just shape them. Perrow notes that elites, in spite of their resources, have limited power to shape organizations because "the complexity of modern organizations makes control difficult" (quoted in DiMaggio & Powell, 1983, p. 157). Universities rival governments in their bureaucratic complexity, a fact which may have made it possible for the Center director to criticize a corporate patron of Letters University's medical school without penalty. Affiliation with medical centers also confers bioethics C&Is with a measure of prestige, and with direct access to biomedical culture and practice, necessary for effective scholarship and teaching.

Slaughter (1997) critiqued the three predominant narratives of curricular change (i.e., the demographic account, the learned disciplines' account, and the market account) in the higher education literature, arguing that faculty members informed by student needs are not the only, or even the prevailing, authors of curriculum. Rather, social movements, the politics of knowledge, and the political economy play major roles in constructing curricula. The preceding account of the establishment of the Center is consistent with Slaughter's assertions. At Letters University, strong student interest was not sufficient to spur faculty initiation of formal curriculum in bioethics, and was not, in fact, even necessary. The later compulsion of medical school faculty to add bioethics to their curriculum appears to have been driven by the need to compete with their peer institutions, to meet granting-agency training requirements, and to address accreditation requirements, rather than by student demand. However, no bioethics professorship, program, or center became a reality until university administrators recognized the potential of institutionalized bioethics to tap into opportunity structures in the world of biomedicine.

The Bioethics Center at Letters University is clearly an example of interstitial organizational emergence, influenced by intermediating organizations including government agencies and accrediting bodies, and crafted by administrators and bioethics professionals in response to institutional competition and new opportunity structures in biomedical technology. By starting up a bioethics center, the university medical center was able to legitimize the pursuit of potentially controversial biomedical research through the explicit or implicit imprimatur of the bioethics center, and also could hope to avoid, or at least foresee, ethical quandaries associated with forthcoming research pursuits.

The entrepreneurial institutional environment into which many bioethics C&Is were born in the 1990s constrains the academic freedom of scholars in the emerging field, caught in the tension between the public-good and academic-capitalism knowledge/learning regimes which coexist in higher education. The Center appears to have been able to establish at least a limited independent critical voice, allegedly by maintaining some distance from university and medical center administration. However, in spite of deliberate attempts to diversify its funding streams, faculty members at the Center still find themselves attentive to the issues that most interest funding providers. The qualitative implications of the academic-capitalism knowledge/learning regime for the relationship of bioethics to established professional stakeholders in the academy will be explored in the chapter 3.

ENDNOTES

1. The National Science Foundation data were provided online by The Center for Measuring University Performance at the University of Florida. This information was retrieved January 21, 2008, from http://mup.asu.edu/ research_data.html.
2. See Mullins (1972) for an account of the stages and social structures involved in the development of new scholarly specialties, as illustrated by the case of molecular biology. Initially, according to Mullins, diverse scholars are drawn together merely out of interest in the same conceptual problem, but over time they organize and distinguish the stylistic and methodological approaches of the new specialty.
3. The results of the survey were retrieved January 21, 2008, from http:// www.apa.udel.edu/apa/governance/committees/medicine/survey/; this survey has not been repeated.
4. This information was retrieved January 21, 2008, from http://mup.asu.edu/ research_data.html.
5. When I conducted my fieldwork in 2002, I was able to identify 18 bioethics C&Is at postsecondary institutions offering bioethics graduate programs. There were other academic bioethics C&Is that did not, at that time, offer affiliated graduate degree programs, such as Stanford University's Center for Biomedical Ethics (which now offers a master's in bioethics).
6. Student interviews did not provide helpful data regarding the establishment, institutionalization, or political environment of the Center, so their responses are not incorporated in this chapter. However, student voices figure prominently in chapter 4, and also appear in chapter 3.
7. The proposal for a Regent/trustee Professorship in Bioethics cited text excerpts from the Liaison Committee on Medical Education's 1989 recommendations, and from the Accreditation Council for Graduate Medical Education's revision of general requirements, approved June 1990. Fuller references may have been provided in appendices to the proposal, which I did not receive. For related discussion in the medical literature, see Dickstein, Erlen, and Erlen, (1991), and Iserson and Stocking (1993a, 1993b).
8. On the 1989 NIH ethics training requirement, and institutional programmatic responses to it, see Eisner (1991).
9. While research on research integrity has not been a central topic in academic bioethics discourse, it represents a funding opportunity to those

interested. The Office of Research Integrity (ORI) in the U.S. Department of Health and Human Services initiated an extramural research program in 2001, offering just over $1 million. In 2005 the allocation for extramural grant funding totaled nearly $2.6 million, and the number of applications was nearly double the 24 submitted in 2001. Information was retrieved January 21, 2008, from http://ori.dhhs.gov/research/extra/award.shtml. For an account of the development of ORI and its performance as a boundary organization, see Guston (2000), chapter 4.

10. While not described as part of its fundraising strategy, the report also noted discussion at the Center about creating a master's program in bioethics, observing the existence of "an enormous appetite for expanding teaching in the area of bioethics" throughout the university.

11. A faculty position announcement appearing in the Center's newsletter specifies, "ideal candidates must qualify for a faculty appointment in one of the departments of the University's graduate or professional schools." Nonmedical faculty members at the Center may be better able to secure employment stability through joint appointments in departments outside the medical school.

12. Indeed, prominent medical professors have been terminated for taking stances contrary to pharmaceutical sponsors of university research. See for example Washburn's (2005) discussion of the cases of Nancy Olivieri, David Healy, and James Kahn.

13. The three existing societies were the Society for Health and Human Values, the Society for Bioethics Consultation, and the American Association of Bioethics. The merger was organized by representatives of all three associations, voted on via mail ballot by the full memberships, and overwhelmingly approved. This information was retrieved April 11, 2008, from http://www.asbh.org/about/history/index.html.

CHAPTER 3

THE ACADEMIC BIOETHICS JURISDICTION: A STRAINED DIALOGUE AMONG DISCIPLINES

This chapter continues the case study analysis of the Center, examining how it seeks to establish its identity and academic credibility within the Letters University community and beyond. As mentioned previously, the compatibility of an academic center's mission with the university's goals, mission, and portfolio of academic programs is critical to academic credibility. Hence, the center must "represent a logical initiative within the university's overall research program" (Stahler & Tash, 1994, p. 550). Here, I explore how faculty and student perceptions of bioethics, and the relationships between the Bioethics Center and other constituencies on and off campus reflect the ongoing construction of an academic and professional jurisdiction for bioethics in Abbott's professional worksite arena, in this case, the existing organizational field of professions and academic disciplines represented at Letter University.

Academic professionals legitimate the work of their profession, in the context of larger societal values, by demonstrating "the rigor, the clarity, and the scientifically logical character" of that work (Abbott, 1988, p. 54). The academic sector of a profession serves the three functions of legitimation, research, and instruction, and the extent to which academic professionals succeed at these functions affects the susceptibility of the jurisdiction to outside intrusion. Abbott observed that "medicine's recent narrowing of its legitimation to science and technology has proved dangerous, since late-twentieth-century cultural values increasingly conceptualize health as quality of life" (p. 54), which he argued contributed to the "humanistic" legitimacy supporting the rise of clinical pastoral education in the 1970s. From the short-term perspective of 1988, he speculates about whether this is a passing phenomenon. I argue that this humanistic legitimacy contributed to the formation of the field of bioethics. However, bioethics has also developed in a period when the professional class has increasingly moved away from an ideology of *social trustee professionalism*, grounded in the ideal of service to the public good, and toward an ideology of *expert professionalism*, legitimated simply by specialized authority over a defined area of formal knowledge (Brint, 1994).

This shift, combined with a more competitive professional marketplace resulting from a growing professional class, has rendered professions more entrepreneurial in nature. The context of these shifting professional ideologies poses a particular challenge to the legitimation of bioethics. While the field originated in the ideal of service, it faces pressures to seek legitimacy through expert professionalism. Furthermore, the biomedical professions that bioethics seeks to influence have become heavily invested in the ideology of expert professionalism.

In Abbott's terms, bioethics can be viewed as having achieved a weak advisory jurisdiction settlement with the strong jurisdiction of medicine, where the powerful profession permits the weaker professional group to serve in a limited advisory capacity. Public claims are particularly important to sustaining an advisory jurisdiction. Abbott (1988) remarked, "It is inconceivable that such [advisory] claims could endure without strong public support. Thus the clergy's practical invasion of hospitals originated as much in a public feeling of the hospitals' inhumanity as in the clergy's own jurisdictional claims" (p. 76). Similarly, the public image of bioethics, and public concern about the ethical implications of biomedical technology, are important to the legitimacy of the bioethics jurisdiction.

Evans (1998, 2002) described how bioethics worked out a weak advisory jurisdiction with life scientists in the moral debate on human genetic engineering, with the result that the state and other bureaucratic institutions became the mediators between science and the public, limiting the scope of the debate. Biologists first succeeded in narrowing debate about interspecies DNA transfer to the issue of safety, effectively excluding the question of whether such DNA transfer should be pursued at all. By circumscribing the debate in this way, biologists kept decision making about genetic engineering within their jurisdiction, because of the technical expertise required to assess biosafety. Life scientists were also successful in institutionalizing the government advisory committee (beginning with the Recombinant DNA Advisory Committee in 1976) as the standard treatment for addressing the social implications of the life sciences, a remedy far preferable to these experts than uncontrollable public outcry leading to restrictive regulation.

An interdisciplinary group of academics (i.e., bioethicists) began to appeal to the public to challenge the jurisdiction of life scientists, arguing that human genetic engineering had the potential to be dehumanizing, and should be limited. However, the audience for bioethics shifted from the public to the state, as state control of science started to become institutionalized. Bioethics commissions became the state's mediator between scientists and the public.

The bureaucratic context and form of these commissions promoted rationalization of the expertise of bioethics, providing the state with a much-needed legitimate means for ethical decision making. Thus, a secular inference scheme was created in bioethics to appeal to the state and compete with the theology jurisdiction. For example, bioethics provided semantically narrow, neutral, universalistic ends (such as beneficence and justice) which were not grounded in any particular religious tradition, and that policymakers could use to make decisions on behalf of "we the people."

Meanwhile, life scientists successfully continued to restrict the scope of the debate on the morality of tasks in their jurisdiction. For example, they limited the scope of the National Commission for the Protection of Human Subjects of Biomedical and Behavioral Research to the subject of human experimentation only (instead of the topically broader "National Commission on Health Science and Society" advocated by Senators Mondale and Kennedy) and established the commission as advisory rather than regulatory (see Evans, 1998, pp. 172–173). Evans (1998) concluded that bioethics is working out a jurisdictional settlement in an advisory capacity to the biologists' jurisdiction, which retains decision-making power. Furthermore, the rationalized, narrow, secular approach that bioethics has largely chosen to adopt is a function of how well that approach maximizes resource acquisition for the profession, strengthening jurisdiction through cultural production of ethical approaches for the state to use that are acceptable to the biology jurisdiction.

What sort of jurisdictional claims are being made for bioethics in the academic workplace? This chapter examines the boundaries of the bioethics jurisdiction as it manifests in the Bioethics Center at Letters University. It begins by developing an account of the self-described identity of bioethicists from interviews with 13 Center faculty members and 11 students enrolled in the Master's in Bioethics program at Letters.[1] Then, it examines the relations the Center has with other key constituencies on and off campus, including academic medicine, nursing, philosophy, and social science departments, as well as IRBs, nonprofit foundations, and corporations. To study these relationships, I not only draw upon my interviews and Center documents, but also supplement my analysis with accounts from the bioethics literature that relate my findings to the broader universe of bioethics in the United States.

THE AMBIVALENT IDENTITY OF BIOETHICISTS

Science reporter Nell Boyce (2002) wrote that the general public views bioethicists as "having a public service role akin to that of a journalist, a government official, or a judge." (p. 17). The opinions of bioethicists are regularly sought by journalists, Congress, and presidential advisory committees, as current events present new ethical challenges, or new instances of perennial challenges. However, as Boyce explained,

> Many scholars reject the title 'bioethicist' because of its 'secular priest' connotation, despite the fact that they work at bioethics centers, write for bioethics journals, and attend bioethics conferences. If the diverse community of philosophers, lawyers, scientists, physicians, and others who study bioethics can't even agree on what to call themselves, how will they reach consensus on their proper public role and their corresponding obligations regarding conflicts of interest? (p. 17)

The diversity of the field and its reluctance to don the mantle of moral authority suggest that this professional group has not yet figured out its desired role and place in the system of professions. A 2001 member survey conducted by the American Society for Bioethics and Humanities (ASBH, 2001a) found that society members were indeed concerned about the professional identity of the field. Preliminary analysis of the survey responses published in the society newsletter summarized members' diverse concerns related to the public image and state of the field:

> Many [survey] respondents specifically stated that the ASBH should develop professional certification standards and accountability measures for clinical ethics consultation. Others wanted ASBH to determine whether there are too many doctoral and master's degree programs in bioethics or to take a leadership role in developing public policy. Some respondents stated an interest in developing the public image of ASBH and the field, expressing concern about the credibility of the field in the public eye. (Gordon, 2002, p. 6)

My case study of the Bioethics Center at Letters University aimed in part to examine faculty members' perceptions of themselves, their field, and the identity of bioethicists. During my visit to the Center I asked faculty members several questions about the direction and identity of the field and its practitioners: What are the goals, or what is the task, of bioethics? What is the biggest challenge facing bioethics, and what progress has been made with regard to this challenge? What characterizes a "real" bioethicist, and do you consider yourself to be one? Is bioethics a field, an emerging discipline, an emerging profession, or something else entirely? The responses of Center faculty and students as a whole, suggest that the identity of bioethics is ambiguous and still developing, and identify several likely obstacles to the cohering of a distinct professional identity in bioethics.

To begin with, some of the faculty I spoke with explicitly or implicitly contested the very idea that bioethics is or can be a singular, easily defined enterprise. One respondent chided the pervasive "tendency to essentialize bioethics," explaining that the term signifies a whole range of activities, directed toward a large set of audiences; there is religious bioethics, and there is secular bioethics—"the diversity amazes." As one of her colleagues echoed, "there is no [one] thing called bioethics."

When asked about the goals or task of bioethics, several faculty members described tasks revolving around (often public) dialogue facilitation, and they were emphatic that it is absolutely not the goal of bioethics to preach or provide authoritative answers. One respondent stated that the goal of the field is to "provide information and skills to let the right questions be asked;" another reported that the goal is to "provide guidance" on large, difficult, and long-lasting issues. One of their colleagues said bioethics' task is to "make *translations* [italics added] between experts and technology and the people that need to use it." Two faculty members expressed the goal of bioethics in terms of bringing *perspective* to stakeholders grappling with bioethical issues. One of these two said that the goal is "to bring multiple perspectives to difficult issues," while the other explained that it is "to help policymakers and the public think through the moral implications of biomedical practice and research, to give them some ironic distance." Another respondent noted that in the United States, bioethics provides a "common language" for dealing with biomedical advances and their social milieu, a language that is compatible with other viewpoints in society.

One respondent asserted that "bioethics is a conversation, not an answer," that the purpose is not for bioethicists to give judgments about right and wrong, "because once you do that you're just another opinion." Rather, the task of bioethics is "to facilitate the public conversation, to say, 'look, we've thought about this a little longer than you have and we know a little more about the background of it, and perhaps even the science of it, and so, having thought about it a little longer and a little harder, let us tell you what we think the issues are.'" Frequently, she explained, the general public does not grasp what the really critical issues are. She provided a striking illustration, worth quoting at length:

> It's taken me a long time to try to get people to see that they're thinking about the wrong issue in genetic discrimination...If you give them [the average person] the kind of insurance line, which is that "genetic mutations are a risk factor just like smoking or any other risk factor, or preexisting condition, and so what's wrong if we give a genetic test to everybody equally and adjust insurance rates based on that, why is that discrimination, when that's what insurance is supposed to do: to take people who are more at risk and people who are less at risk and generalize that across the population so that everyone subsidizes risk, but in addition charge people who have identifiable greater risk more?" And there are a number of answers to that, but one that people don't normally recognize is that the discrimination will come not because they're not giving genetic tests fairly, but because the genetic tests we have were developed because of research on specific populations: Amish, Mormons, Ashkenazi Jews, Icelanders, Finns, French Canadians...they're fairly genetically homogenous, so they make good genetic subjects.
>
> What that ends up meaning is that we have genetic tests first for those diseases that these populations have and therefore, even if we give those tests to everybody, they will disproportionately impact those populations..., the very populations that have given their time and effort to geneticists to allow them to further the field. Meanwhile, other groups who have not been under the

geneticist's microscope may have other diseases, but we don't have tests for them yet. And so those people's genetic weaknesses will not be identified, while French Canadians' will be. And therefore, even if you give the genetic tests we have to everybody, they're still discriminatory. Well, you're [i.e., the average person] not going to think of that just thinking about, "Are genetic tests discriminatory?"...*That's one of the roles of the bioethicist to deeply understand the issues and then be kind of an emcee for the public conversation that needs to come from that* [italics added].

Two respondents ascribed more active goals to bioethics than the facilitation of conversation. The first described the task of bioethics in terms of general academic functions: to create knowledge, educate the public, and perform outreach. She explained that outreach "can include trying to actually bring about practical things that can help people in clinical settings to improving corporations to trying to bring about public-policy changes." One of her colleagues took a more radical stance, saying that the goal is

stopping the corporate biomedical machine, a little bit, making them pause before they just plow forward...I don't think that the medical system should be for-profit...I think in some ways bioethicists are pretty radical thinkers, and while some of them think that it's okay to have a for-profit health care system, I do think that they have a critical role. That's one of the broad goals. They're really charged with raising questions...and they try to do it at the forefront of decisions. And that, in itself, is pretty admirable, even if they don't necessarily do it early enough.

A third faculty member, however, viewed bioethics as failing at reformism. She commented that in the 1970s, bioethics was attractive because it was "largely toothless." The consumer rights movement, which was calling for the institution of patient advocates, was far more confrontational than bioethics, which presented a more "gentle social solution" that was "in league with, not against, medical authority." However, she still saw considerable value in bioethics, asserting that "the world is a much better place with bioethics than without it."

After asking faculty respondents about the goals of bioethics, I asked, "What is the biggest challenge for bioethics?" Their responses addressed the themes of legitimacy, identity, the development and use of critical skills, and structural issues. A few faculty members expressed concerns related to the identity of the field. One respondent viewed the biggest challenge for bioethics as "defining its spheres of influence." She elaborated,

> For example, is it clinical, is it research, is it theoretical? Is it philosophic or sociologic or anthropologic? And the answer to all of those can be yes, but I think we have to say up front "here's what we do, and here's what we don't do."...As with any discipline, those who aren't as well entrenched in it may not recognize all the spheres, and may establish a hierarchy of the spheres and then say, "here's what I do, and here's what you do."

She indicated that there were few individuals she considers to be bioethicists, and noted that one of those individuals "crosses boundaries between the spheres very well." One of her colleagues commented that the field is "sort of floundering around," trying to figure out what it is, and that "it's a lot of different things." She was not herself uncomfortable with this ambiguous identity, but thought it might be uncomfortable for others desiring bioethics to be a legitimate field.

Other respondents' concerns reflected the sense that bioethics has a problematic identity, in terms of what the key competencies that characterize bioethics are. One faculty member felt that the biggest challenge for the field was to bridge the gap between philosophical discourse and practice, observing that it is very difficult to develop both skills. One of her colleagues asserted that bioethics is "philosophically thin"—that the field has no canon, making training difficult. Taking a different perspective, another faculty member argued that bioethics needs to do a better job of using social psychology and economics in bioethics work.

One faculty member said that what most required attention was the improvement of research skills, "if bioethics is to emerge in a holistic fashion;" another declared there is a lack of good cross-cultural bioethics research. According to one respondent, the field needs to establish a "solid intellectual and technical base," such as instruments, scales, and benchmarks for scoring how well legislation and policy addresses bioethical issues and goals. From these observations, it would appear that the field as a whole is not currently doing an effective job of bridging the various "spheres" putatively embraced by bioethics.

Several faculty members identified structural issues as the biggest challenges for bioethics. The Center director felt that the biggest challenge is the "traditional home" of bioethics in academic medical centers, which, as a whole, represent "an institution in flux," likely leaving bioethics centers vulnerable. One faculty member stated that what most requires attention in the field of bioethics is the establishment of independence from the press and "high-profile issues of the day." Another respondent asserted that bioethics is not a vibrant, growing field, but just seems large and powerful because of its media presence. The Center director also cited diversity as an important structural challenge, noting that "in bioethics it's important because diversity helps give you more perspectives on ethics issues." She said that while there are more women in the field now, there are few ethnic minorities, a situation she attributed largely to the fact that there are few ethnic minority academics in health care fields, or in universities generally.

Another challenge for the field of bioethics is what I call "biotechnofetishism," an excessive fascination with or regard for sensational biotechnology, that overshadows the chronic, systemic unglamorous challenges of biomedicine. One of my faculty respondents reflected,

> The tension between following the technology and trying to be useful to health care is another challenge for bioethics...There's always a pull to the new interesting technology, and yet most people aren't affected by that. I mean they are, to talk about it socially, but if you're really in the health care system you still want to know why you have to wait 4 hours before they see you in the ER, and why there are so many uninsured people, and why are nursing homes a mess. That's a tension. It's not insoluble, *it's just a challenge to kind of cover the areas and not get seduced by the fancy*

cool technology all day long [italics added]…I write about a lot of fancy cool technology stuff (transplants, stem cells, human experimentation), but I do try to write periodically about nursing homes, rehab medicine, managed care, and health insurance…*I don't want to forget to hit the mundane problems of health care for all the gee-whiz problems. The field needs to struggle with both* [italics added].

I found that the philosophers among the Center faculty were the most likely to "get seduced by the fancy cool technology," whereas the clinicians among the faculty were more likely to grapple with issues in clinical practice and research (e.g., end-of-life care, informed consent), as evidenced by their recent publications listed in the Center's annual report.

The Identity of Bioethicists:
Multidisciplinary Quasi-Professionals?

Is bioethics a discipline? Is it a profession? Is it becoming one or both of these things? What is a bioethicist, exactly? During my site visit to the Center, I asked both faculty and students whether they considered bioethics to be a profession, and what they felt distinguished someone as a bioethicist. I also asked faculty whether bioethics is a discipline, or is becoming one.

Students were familiar with the debate on the identity of bioethicists, and a few were more comfortable than faculty in calling themselves bioethicists—in spite of the fact that faculty would be more readily identified as bioethicists, given their positions in university bioethics centers, programs, and departments. Furthermore, several faculty members expressed dismay that some Master's in Bioethics (MBE) students would call themselves bioethicists. One student remarked that perhaps philosophers and theologians were more likely to adopt the title of bioethicist than scholars and practitioners from other disciplinary backgrounds were. Another student, who was not comfortable calling herself a bioethicist, stated that she would rather be called a student of bioethics, in the same way that one would prefer being called a painter rather than an artist—i.e., to convey a more modest level of mastery, and/or less pretentiousness. One of her peers, who also was reluctant to call herself a bioethicist, quipped that students would come out of the program "with just enough knowledge to be dangerous."

Several students pointed to the combination of a related formal knowledge base (including having a foundation in a traditional discipline such as medicine, law, or philosophy, and being versed in the tools and rigorous analysis typical of the field) and appropriate firsthand experience (e.g., work a clinical setting, or conducting research) as the hallmarks of a bioethicist. One student speculated that bona fide bioethicists probably have degrees in philosophy, but that it was on-the-job learning and a keen interest in ethics that mattered; bioethicists "don't need a sheepskin," because there is no defined beaten path for work in bioethics. Another student asserted that *self-developed* expertise, through education, clinical experience, and research, was the signifying characteristic of bioethicists. One of her classmates made a slightly different distinction, pointing to expertise, *relative* to the lack of bioethical expertise of other professionals in a particular setting, as the defining feature; this student felt she would be comfortable calling herself a bioethicist in the clinical arena. One student questioned whether one could be a bioethicist without being a nurse, doctor, or some other health professional. When it comes to clinical ethics, she argued, bioethicists need a thorough knowledge of the human body. With regard to lawyer bioethicists, she felt that these individuals would really need to engage in close relationships with health care providers.

I asked faculty members at the Center what signifies a "real" bioethicist, given the lack of a defining credential. I also asked how comfortable they were being designated a bioethicists themselves. Most respondents reported that they either were uncomfortable with the title, or did not consider themselves to be bioethicists, instead primarily identifying themselves according to the disciplines in which they were credentialed (e.g., philosopher).

Two faculty members were only "sometimes" uncomfortable, and two were comfortable calling themselves bioethicists, although one qualified her response in terms of her definition of bioethicist, that is, having the organizational title or appointment, professor of bioethics. The tendency to feel uncomfortable with the title of bioethicist, I was told, is rooted in a strong skepticism of the very idea of moral authority. As one faculty member put it, she is "troubled by the idea that somebody knows right from wrong, and has special skills in that area. The name [bioethicist] sounds like that." Several of her colleagues reported discomfort with being called a bioethicist, particularly in situations "where people want to be told what to do."

As mentioned earlier, several faculty members expressed dismay that students in the MBE program comfortably called themselves bioethicists, because these academicians believed a master's degree could not provide sufficient training to prepare a student for that role. One faculty member reconciled the disconnect between faculty and student with regard to being comfortable with the designation of bioethicist, explaining,

> Bioethics started out as a field where everybody could come in regardless of discipline or background, and it's starting to professionalize, and it hasn't figured out what to do about that. There's a lot of tension about "Am I a bioethicist, or am I just a sociologist that studies bioethics?" Younger people have no doubt what they are; if we give them a degree in bioethics, they're bioethicists. They're going to say, "Of course I'm a bioethicist, I have a degree in it."…It's going to change the issue of "Can it be interdisciplinary?" to, "Is it just one more subject area?"—it's not pulling people together [from different disciplines], it's just full of people who are bioethicists, that's what they do.

As the younger generation of bioethicists, now being trained in full-fledged bioethics graduate programs, begins to populate faculty positions and participate in annual ASBH meetings, I anticipate that that programs will exhibit all three aforementioned forms of isomorphism, generating a bioethics canon and a standard curriculum, and to some extent losing their individual niche orientations which resulted from their different agent advocates, institutional cradles, parent departments, and development strategies. A few bioethics PhD programs have been established, and more are anticipated, suggesting a continuation of recent growth in the field. If my prediction is accurate, we can expect the field to develop a more distinct, more homogeneous identity, provided that a stable jurisdiction is staked out in the professional arena, and that bioethics continues to grow. If not, bioethics may recede back primarily to the function of clinical ethics education, with a marginal academic institutional presence.

Faculty respondents' various accounts of what signifies a "real" bioethicist focus on social relations among experts, which bind together the interdisciplinary field that has been colonized by professionals with an array of credentials, experiences, and skill sets. Two faculty members did point to training in moral philosophy as a defining characteristic of bioethicists (neither of them held a philosophy degree, and neither considered herself a bioethicist). Some respondents gave seemingly tautological answers. One faculty member asserted that the *work* of bioethicists, that is, research and teaching in bioethics, is their defining characteristic; similarly, another respondent stated that bioethicists "produce things other bioethicists recognize as bioethics." She continued,

> There are some things that give you clues as to who the bioethicists are: they work in [centers] like this, they belong to bioethical organizations, they publish in journals about ethical issues. And the borders are murky, but in that sort of critical mass, people are either accepted or not accepted based on their works, and so it's a consensus kind of thing…Journal articles and public conversation about bioethical issues, those are the products that brand one as a bioethicist.

Acceptance into the fold of bioethics, suggested my respondents, was contingent upon such characteristics as publishing in the right places and having appropriate clinical experience (e.g., as a health care provider, chaplain, hospital social worker, or ethics consultant). One faculty member stated that MBE recipients were "in the club, in some sense." Having received this "trial membership," MBE holders are then subject to "hazing rituals," she joked, in the form of the requirement to publish, and otherwise remaining current with and engaging in ongoing activity in the field.

Most faculty members felt that bioethics is multidisciplinary or inter-disciplinary, a field rather than a discipline. One respondent did remark that bioethics is "emerging towards a discipline," as evidenced by a "common lingo" and "core [topical] interests." One of her colleagues argued that bioethics is "a subject area, not a discipline," and felt that students should be trained in a discipline at the graduate level. Similarly, another respondent contended that bioethics has "no independent intel-lectual moorings," and is a multidisciplinary or interdisciplinary activ-ity in which scholars address "a domain of questions from disciplinary stances." Yet another respondent felt that bioethics was multidisciplinary, and not strongly interdisciplinary; it constitutes "an area of inquiry and action," with a core set of questions and issues, and "latent themes."

Faculty members at the Center were also rather hesitant to identify bioethics as a profession. As cited previously, one respondent com-mented that bioethics is "starting to professionalize, and it hasn't figured out what to do about that," referring to the divergence between faculty and students in designating themselves as bioethicists. One of her col-leagues felt that bioethics is not a distinct profession, though it might become one; it currently lacks a location or arena of practice, she said. Another faculty respondent noted that bioethics does exhibit some of the hallmarks of a profession, such as the consolidation in 1998 of three bioethics societies into one (as mentioned in chapter 2)—the American Society for Bioethics and Humanities (ASBH)—and the proliferation of master's degree programs in bioethics.

Another prima facie hallmark of bioethics professionalization is the creation of the Task Force on Standards for Bioethics Consultation (referred to hereafter as the Task Force), which published the report *Core Competencies for Health Ethics Consultation* (1998).[2] However, the report itself implies reluctance on the part of the Task Force to promote the professionalization of bioethicists, at least as clinical ethics consultants. The text recommends that the report be used as *voluntary* guidelines, rejects certification to perform ethics consultation, and similarly rejects granting accreditation via educational programs. As we are about to see, the reasons given in the report for limiting the formalization of its recommendations underscore a prevailing aversion in bioethics to stake a claim of moral authority, as well as an ambivalence toward the embracing the ideology of expert professionalism.

The report (ASBH, 1998) noted that the Task Force "does not intend for its report [to be] used to establish a legal national standard for competence" (p. 31) for several reasons:

- Firstly, certification is rejected because it "increases the risk of displacing providers and patients as the primary moral decision makers at the bedside," which "could encourage the type of authoritarian approach to ethics consultation the Task Force has rejected" (p. 31).
- Secondly, the report argued, "It is important that consultants have relevant competencies, not that they come from some particular professional or academic field," and the Task Force does not want to "undermine disciplinary diversity" that "leads to a more balanced understanding of competencies" (p. 31).
- Thirdly, certification "could lead to the institutionalization of a particular substantive view of morality," of ethical approaches, or of conceptions of the relative importance of consultation skills (p. 31).
- Fourthly, the report stated, "it is unlikely at this time that a sufficiently reliable test could be developed to measure the required competencies" (p. 32).

These justifications suggest an aversion not only to claiming moral authority, but also to defining the jurisdiction in terms of professional expertise. Rather, ethics consultants exist to advise other professionals and persons who actually make health care decisions, and may competently do so by drawing on overlapping but variable skill sets.

One of my faculty respondents at the Center asserted, "Bioethicists are not a professional group," but a collection of people with a common substantive interest, and with different backgrounds. Rather, the "role" of bioethicist, she contended, is one that has been created by some of the leading personalities in the field, who have identified themselves as bioethicists.

Rather ironically, it appears that acceptance of one's work by reluctantly self-identifying bioethicists is the key to community membership. The field appears to be largely self-defining, wary of the treacherous moral-authority connotations of its designation, reluctant to define a standard expertise, and maintaining itself in recruitment mode, rather than drawing pointed distinctions between bona fide members and mere pretenders. Control of membership is exercised simply through the gate-keeping of publishing and hiring.

At the same time, the established generation of bioethicists, who identify themselves primarily in terms of their respective original disciplines, are also wary of institutionalizing bioethics credentials, because the field does not possess the distinctive methodology, rigor, theory, and canon that together characterize "pure" academic disciplines. There is also widespread aspiration in the bioethics community to preserve the multidisciplinary nature of the field, and concern that this aim will be difficult to achieve as the field becomes more institutionalized. Nonetheless, bioethics master's programs have multiplied, and graduates are far less reluctant than their professors to adopt the title of bioethicist.

The inherent multidisciplinarity of the field of bioethics is a characteristic embraced by its membership as an approach to addressing the complex relationship between biomedicine and society, but one that makes it difficult for the field to formulate, agree upon, or convey a strong professional unity and outward identity, distinct from the various disciplines it draws upon. As former president of ASBH John Lantos (2002) observed in the society newsletter,

> As a field we've carefully, although sometimes passively, avoided narrow specialization. Though there are now both undergraduate

and graduate programs in bioethics, there are no standard curricula, no standard training, and no uniform certification in bioethics. Instead, we've let a thousand flowers bloom, trying to remain true to our interdisciplinary roots—to honor theology, philosophy, literature, film, disability studies, cultural studies, feminist studies, queer studies, clinical ethics, virtue ethics, principalism [*sic*] and casuistry, economics, law, history, the social sciences—while at the same time trying to bring some coherence to our inquiries, some relevance to the fact that the work we are doing is work in **bioethics** [emphasis original] rather than one of those other fields. (p. 2)

It seems that bioethicists are consciously avoiding becoming members of either a discipline or a profession, although there are pressures both pushing them toward, and pulling them away from both of these categories. The need for legitimacy and financial resources push them toward claiming expertise, defining that expertise, packaging and marketing it in master's degree programs, and seeking research and training grants to expand that expertise. On the other hand, the reluctance to claim moral authority, and the entrenched power of the professions and institutions which bioethicists advise, restrict the influence and jurisdictional claim of bioethics.

THE RELATIONSHIPS BETWEEN THE BIOETHICS CENTER AND OTHER JURISDICTIONS

Having presented my findings concerning the self-perceptions of current and fledgling bioethicists at the Center, I now turn to the boundaries between bioethics and other jurisdictions, both on and off the Letters University campus. The jurisdictions discussed here are not exhaustive, but rather include several that emerged from the data as being of particular interest. My analysis draws on faculty interviews, student interviews, and Bioethics Center documents I collected during my site visit, and is supplemented with accounts from the bioethics literature that relate my findings to the broader universe of bioethics in the United States. One of the Center documents I used, referred to here as the Jenkins Report, was an evaluation of the Center, prepared by a consultant hired by the Center, in anticipation of a periodic review of the Center by Letters University.

Here, I consider the relationship of bioethics to academic medicine, academic nursing, philosophy, sociology, institutional review boards, hospital ethics committees, research integrity training, nonprofit foundations, corporations, the public-policy arena, and the community audience to which academic outreach is directed.

Bioethics and Academic Medicine

Given that the traditional home of bioethics is in academic medical centers, what dynamics characterize the relationship between the two? As we have already seen, the medical school dean (eventually), the director of medical education, and medical students, as well as the director of the Biotechnology Institute, all saw value in bringing bioethics to Letters. The Center director reported that there was no resistance to the Center developing a role in medical ethics education at the medical school. She admitted being a little apprehensive at first, concerned about potential perceptions that the Center was imposing on the medical school, but observed that "people have been open."

At the time of my site visit, several medical school faculty were affiliated with the Center. One of my faculty respondents, who had extensive experience in clinical settings, both as a consultant and in conducting ethics research, reported that the relationship between the Center and the university medical center was not contentious; she felt welcome at the university hospital. She observed that, "Clinical staff are often frustrated because they feel their jobs are getting microfocused by bureaucratic structures, and they worry about losing the big picture." She provides an outlet for their frustrations, and they are "happy someone is taking a broader approach" to dealing with health care delivery issues. In fact, she even described an occasion in which department heads had invited her to attend the planning meeting for a clinical trial. "They don't want to lose the human side [of medical work]," she explained, reflecting that clinicians felt constrained by the system, and wanted to feel they were justified, in a documentable way, for taking "heroic levels of action."

While relations between the Center and the university medical center were reported to be cordial, I found no evidence that the Center was actively working to shape clinical policy and practice within the University. When asked whether the Center makes efforts to shape Letters University policy, one respondent, who works out of the medical center, said, "not at the [University] hospital…[the Center] doesn't pressure the hospital." She reported that there was not much "overlap" between the Center and the university hospital, noting that "a larger linkage is assumed, but they operate separately."

Bioethics and Nursing

At the time of my visit, two nursing professionals were faculty affiliates of the Center. One of these faculty members, also a faculty member in the nursing school, told me that the local nursing ethics community was "thrilled" about the Center; one of them sat on the Center's advisory board. She felt that bioethics is beneficial to nursing; it raises "big questions for nursing at appropriate times," she said, and has, for example, improved the use of advance directives in clinical practice. There was a good relationship, she claimed, between nursing and the Center, commenting that "nurses are hungry for bioethics." She pointed out that several graduate-level nursing students were taking courses offered by the Center, and that several nurses were enrolled in the MBE program.

However, she contended, nursing ethics education needs to be very applicable to daily practice, and it would take a lot of time to integrate appropriate training into the Center's curriculum. She described her role at the Center as bringing a clinical perspective when appropriate, and stated that she did not feel the need to influence the Center. She also explained that nursing ethics takes a different approach than medical ethics; while medical ethics pragmatically addresses "what doctors should do," nursing ethics is based on a theoretical framework, and employs several ethical models.

The MBE students I interviewed who came from the nursing school were a little more critical of the relationship between nursing and bioethics, both on the Letters campus and in general. One student acknowledged that while bioethics is broad, and nursing is just one aspect of it, she viewed bioethics as a "medical" field, where the focus is on physicians and acute health issues, instead of on pervasive health care delivery issues. She remarked that she and one of her nursing peers in the MBE program often left the bioethics classroom "exasperated," because of the difficulty of making their classmates "think outside their area of comfort, which is all about high-tech medicine," and engage with less exciting issues, such as how resources are being allocated to people who are marginalized.

Another MBE student from the nursing school said, "I get the sense sometimes that they [some of her professors at the Center] look at nurses as being less able to contribute." When asked why she thought the contribution of nursing was underappreciated, she pointed to the fact that bioethicists are drawn primarily from the ranks of physicians, attorneys, and PhD holders. She also reluctantly commented, "I think nurses are not always articulate, and they sometimes take a lesser role just because of the [subordinate nature of the] profession, but I think we have a lot to offer." She predicted that as more nurses enter bioethics, they will have a greater influence on the discourse.

This respondent was also critical of the "alleged" connection between the Center and Letters' nursing school. Although the nursing school offers, and advertises, a dual graduate degree in nursing and bioethics, this student claimed that she, along with one of her classmates, encountered articulation problems between the nursing and bioethics programs. Her advisor in the nursing school did not understand how the joint program worked. Similarly, the Jenkins Report (an evaluation of the Center by an independent consultant) noted the lack of a close working relationships between the Center and the nursing school, recounting, "some [Center] faculty suggested [the Center should pursue] stronger ties with the school of nursing."

Bioethics, Philosophy, and the Social Sciences

The Jenkins Report noted that "[Center] faculty expressed concerns about the perceived poor relationship between the Philosophy depart-ment and the Center, though were hesitant to place blame on anyone at the Center." One of my faculty respondents was particularly aware of this issue, due to her experience with Letters University's philosophy depart-ment. She described a disdain, or gulf, between bioethics and philoso-phy. In her view, philosophy champions analytic rigor, while bioethics champions real-world relevance; philosophy, she asserted, "didn't want to get its hands dirty." However, only if bioethics is "philosophically thin," she argued, does philosophy have a serious critique of bioethics. As she noted, PhD programs that have bioethics components tend to be weak in philosophy. One of her colleagues affirmed that philosophy, in contrast to bioethics, was intentionally remote and isolated from the world. Another faculty respondent contended that some philosophers harbor resentment and jealousy toward bioethics, for the attention and resources the field has garnered.

The social sciences have, in general, an ambivalent relationship with bioethics, due to the opposing intellectual orientations of philosophy and social science, and to the social science aversion to claiming normative stances. One faculty member commented that philosophy and sociology are "diametrically opposed methodologically," and she is not sure how the two worlds can be bridged. However, she also cited the importance of balancing philosophy and social science in bioethics, in order to examine what is actually happening in the world, but also to pursue nor-mative analysis. One of her colleagues echoed this sentiment, asserting that the field needs to "mix the normative [analysis] and social science more." One of my respondents claimed that "social scientists hide at ASBH [annual meetings]," explaining that social scientists defend the claim of objectivity, and are uncomfortable supporting normative claims. Another possible reason for the tension between the social sciences and bioethics is that social scientists have waged scathing critiques of bio-ethics (as reviewed in chapter 1), critiques which may not always have been welcome.

Social scientists' participation in bioethics is also reminiscent of natural scientists' participation in public-interest science organizations during the Cold War period, whereby "activist scientists sought new ways to maintain credibility simultaneously as objective scientists and as political actors serving the public good" (Moore, 1996, p. 1593). However, accusations of political and corporate taint in bioethics have made it difficult for social scientists to maintain objective credibility while simultaneously engaging with critique of biomedicine.

Bioethics and IRBs

As mentioned previously in this chapter, one faculty member predicted that research ethics will create enormous demand for bioethicists, and become a primary activity in the field. However, considering the challenges faced by several bioethicists in engaging with IRBs, as evidenced at Letters University and elsewhere, the interface of bioethics and research oversight appears to be a problematic one, at least in competitive research institutions.

Some of my field sources expressed the need for bioethics to increase its engagement with research oversight. The Jenkins Report notes, in a section of the text addressing academic collaboration and community outreach, "Some clinical faculty expressed concern about the lack of a strong presence [by Center faculty] on the [Letters] IRB and strongly recommended [Center] involvement in that Committee." One MBE student envisioned an active role for bioethics in relation to IRBs; observing that the IRB oversight mechanism "isn't right yet," she suggested that it may be up to bioethics to guide IRBs to not merely approve research projects with human participants, but to monitor them more actively.

Several faculty members I spoke with reported having clashes with the IRBs they had previously served on at Letters or elsewhere. One faculty member described her relationship with the IRB she served on at Letters as "contentious," noting that "making too much noise" was discouraged. The IRB, she explained, is an "institutional creature, there to get research done," a bureaucratic organization that promotes form over substance. She was frustrated with the process. She also noted that the

Letters budget for IRBs had quadrupled, that a new chief IRB administrator had been recently appointed, and that a new triage system had been implemented to expedite exemptions. A Center staff member remarked that Center faculty "keep getting kicked off [Letters] IRBs," due to differences of opinion about templates and review procedures, suggesting that the relationship between the Center as a whole and the Letters IRBs was contentious.

Other Center faculty reported serving on IRBs outside Letters, where their experiences varied. One respondent was an IRB member at a medical college elsewhere earlier in her career, but had not served on a Letters IRB. In her experience, there is a lot of tension when a bioethicist sits on an IRB; a bioethicist has different priorities than physicians and scientists do. The IRB is a flawed institution, she said, but no reasonable solution has been offered that is not IRB based.

Another faculty member reported serving on an IRB outside Letters, and had a more positive experience. She reported it to be a "congenial IRB," whose members were unaware of some issues, but otherwise "relatively well informed." It is notable that this particular IRB reviewed mainly social science surveys and protocols, and met only a few times per year. I learned that the institution served by that IRB is not primarily driven by the motivation to earn revenues from large research grants, or to churn out research articles in top-tier journals, but rather had an explicit action-oriented mission, in contrast to the competitive biomedical research environment at Letters.

Center faculty members' experiences as IRB members are not unique to the research oversight culture at Letters. In *The American Prospect*, Carl Elliott (2001) of the University of Minnesota Center for Bioethics described his contentious experience as an IRB member for a psychiatric hospital. Although he invoked international, national, and university guidelines that prohibit placebo-controlled psychiatric clinical trials, and was supported by other IRB members, the board Elliott sat on still approved many such drug trials. He recalls, "Tables were pounded. Faces turned scarlet. Blood pressures soared...The hospital administration eventually dissolved the IRB and reconstituted it with new membership" (p. 18). The

IRB counted among its members a number of hospital psychiatrists who worked with pharmaceutical companies to conduct placebo-controlled drug trials, and who correctly pointed out that the FDA requires evidence from placebo-controlled trials in order for new drugs to be approved, in spite of the opposing ethical guidelines Elliott had just cited.

A recent study found that almost half of academic faculty members who serve on IRBs have had consulting relationships with pharmaceutical companies within the 3 years prior to that service (Campbell et al., 2003). According to bioethicist George Annas, IRBs are under pressure to approve studies in order to bring in desperately needed research dollars (Washburn, 2005). While IRBs constitute a forum in which the expertise of bioethicists is especially relevant, they are often difficult settings for bioethicists to effectively engage in, given the close ties with industry and potential conflicts of interest that apply to many clinical researchers who sit on IRBs alongside the bioethicists.

At Letters, the Bioethics Center had better success engaging in other ways with the university's research oversight structure. The Center developed a research ethics training program, to fulfill training requirements for all investigators at Letters University working with human subjects. One Center staff member said, "We want to get more involved in the educational component [of research oversight]." She also conveyed that the Center had a close relationship with the director of IRBs at Letters, who was reportedly supportive of greater Center involvement in human subject protections activities at Letters, and expressed interest in university participation in the Center's research ethics improvement activities.

The relationship between bioethics and research oversight is a complex one. Many IRB administrators are probably receptive to working with bioethicists in order to guide and shape research ethics through empirical research on research ethics, and through education. The problem arises when high-stakes biomedical researchers serve on IRBs alongside bioethicists. Particularly in competitive biomedical research institutions, the organizations' achievement-oriented culture, aimed at procuring large grants and publishing in high-prestige journals, co-opts the research oversight mechanism. However, IRBs are not always, or even usually, a rubber-stamping operation. I have seen IRBs being accused of excessive caution in the protection of human subjects when faced with protocols involving genetic research with which the board members were unfamiliar, and have heard investigators complain repeatedly about the tremendous challenges of having a protocol approved by all of the multiple IRBs overseeing a multisite study. Bioethicists are undoubtedly welcome members on many IRBs, and may become more welcome as the threat of research malpractice litigation builds, an issue I will address in chapter 5.

Bioethics, Hospital Ethics Committees, and Research Integrity
The relationship between the Center and hospital ethics committees (HECs) and research integrity activities, on the one hand, was less problematic than the Center's relationship to IRBs, on the other hand, insofar as the Center faculty chose to spend their limited time engaging with these areas.

Hospital ethics committees (HECs) are a venue similar to IRBs, where bioethicists may serve as members alongside professionals from other jurisdictions. Like IRBs, HECs have been criticized both for going too far, and for not going far enough, in ensuring that physicians make ethical clinical care decisions. However, unlike IRBs, HECs are rarely a regulatory requirement,[3] and there are no enforced standards for the composition or training of these committees.[4] Nonetheless, the courts have more often than not deferred to HEC decisions as expert decisions. HECs began to grow substantially in number during the 1980s, in response to the Reagan administration's efforts to require life-saving care for all disabled newborns (Rothman, 1991). Several major medical professional associations officially supported the establishment of HECs (Peirce, 2004), which were undoubtedly preferable to hospital administrators and health care providers than government regulation or adjudication of clinical ethics dilemmas in the courts would be. By 1998, more than 90% of U.S. hospitals reported having an HEC (McGee, Spanogle, Caplan, Penny, & Asch, 2002).

A Center newsletter published during the Center's fourth year of operation reported that 4 faculty members at the Center served on

HECs, but none of the faculty members I interviewed during my visit mentioned serving on HECs, or commented on these bodies. I did not see any indication in the literature or during my site visit that the relationship between bioethicists and HECs is particularly contentious. Given the high-stakes funding behind studies reviewed by IRBs (which are populated to a considerable degree by investigators supported by state or corporate sponsors) and relative the lack of such individual conflict of interest among members on HECs, one would expect that bioethicist committee members have had better success at contributing to decision making of HECs than to that of IRBs.

I spoke with one Center faculty member about teaching research integrity training to graduate students in the life sciences at Letters. She and three other Center Faculty members had a team-taught course in research integrity a few years prior to my visit, though she was no longer involved with this activity. She said that they accomplished no real reform in this area, but they had inspired the students to be proactive in ensuring the integrity of their own research. Letters fulfilled the basic federal requirement for research integrity training, she said, but was not taking the innovative approach that some of its peer institutions were pursuing, such as employing simulations. The Center's involvement in such efforts was limited mainly by available faculty time, and not by any desire on the part of university administrators to exclude input from the Center. As of spring 2006, the Center was involved in providing NIH-mandated research integrity training to biomedical postdoctoral fellows at Letters. I did not identify any other explicit involvement of the Center in Letters research integrity activities.

Bioethics and Nonprofit Foundations

As discussed in the previous chapter, the Center director and other faculty regard nonprofit foundations as an important source of funding, as well as a stakeholder in which issues are addressed by academic bioethics. While faculty work is shaped by the priorities of foundations, one faculty member also pointed out that conversely, bioethicists also have some influence on foundation agendas. For example, she asserted that William Stubing, the president of the Greenwall Foundation, is receptive to input from bioethicists, and takes an interest in innovative work. The Greenwall Foundation only funds grants in the program areas of bioethics, arts, and humanities, and disbursed over $2.5 million in grant funding for bioethics in 2008. As its Web site stated in 2005, "The Foundation is especially interested in supporting pilot projects and the work of junior investigators, and it is prepared to address issues regarded by some as sensitive or potentially controversial" (Greenwall Foundation, 2005).[5] Such controversial issues have included assisted suicide; one of the Greenwall Foundation's grantees is the organization formerly known as Choice in Dying (now called Partnership for Caring), which has been labeled a "right to die" group.[6]

Bioethics and Corporations

Bioethicist Carl Elliott is outspoken about his discomfort with accepting funding from corporations. To him, it "seems too much like bribery. If it's not bribery, it becomes the perception of bribery" (quoted in Boyce & Kaplan, 2001). Laurie Zoloth (2001), past president of the American Society for Bioethics & Humanities (ASBH), argued that rather than financial inducement, the real seduction for bioethicists is the prestige and attention conferred by corporate consultancies—and, as she notes, "For this sort of social capital, there are no restrictions, and no regulation," in contrast to existing rules addressing financial conflict of interest (p. 17). Arthur Caplan, first president of the former American Association of Bioethics, on the other hand, felt that conflicts of interest can be managed, and that there are real benefits stemming from bioethicists' engagement with corporations, "where much of the controversial, cutting-edge science gets done" (Boyce & Kaplan, 2001).

Finding fault with corporate activities often comes at a price. For example, pharmaceutical manufacturer Eli Lilly halted its annual donation to the Hastings Center after the Hastings Center's journal published articles critical of Lilly's drug Prozac (Boyce & Kaplan, 2001).

As discussed in the previous chapter, several faculty members at the Center emphasized the importance of limiting the amount of corporate funding accepted, and of taking steps to minimize conflict of interest and to protect academic freedom. However, two respondents also affirmed the value of engaging with corporations to achieve bioethical objectives. One respondent remarked that she found the success of some environmental consultants in the "greening" of corporations inspiring, and that, by working with corporations, bioethicists could have a similar impact. She also asserted that it is difficult for bioethics to shape public policy directly, and she thought that one might have greater influence by working with corporations.

In fact, corporations sometimes seek out bioethicists; one faculty member was approached by a biotechnology company after one of its executives saw her quoted in a story in a weekly news periodical. Another faculty respondent noted that in order to even pursue scholarship on some topics, such as embryonic stem cell (ESC) research, one almost *needs* to work with corporations; at the time of my field visit, there was very little government funding offered to conduct ESC research or investigate the ethics of it. Virtually the only way to obtain access to ESC research or funding to conduct ethical analysis of it, she said, was to "find a way to work with the people who do it [ESC research]." However, she also pointed out, as did one of her colleagues, that openly criticizing corporate biotechnology, such as pharmaceutical corporations, all but excludes any possibility of working with corporations in that subsector.

Bioethics and Public Policy
As mentioned previously, one Center faculty member asserted that it is a significant challenge for bioethics to influence public policy. One of her colleagues made the distinction that while it is difficult to impact federal-level policy, "the future of bioethics," she said, is in state-level policy. She cited the example of human cloning: there have been several bills in Congress to ban cloning, but none were passed into law. Furthermore, a cloning ban could turn out to be unconstitutional, as a violation of First Amendment rights to freedom of speech (see, e.g., Hsu, 1999). Conversely, at the time of this writing, 12 states had enacted statutes banning reproductive and/or therapeutic human cloning, and 2 other

states have legislated prohibitions against the use of government funds for human cloning (NCSL, 2008).[7] My respondent also noted, however, that there is less prestige and visibility associated with state-level policy work, suggesting that pursuing it will have less positive impact on the institutional legitimacy and public reputation of bioethics than pursuing federal-level policy work.

I will return to the issue of bioethics' interaction with federal-level policy in chapter 6, where I examine the interaction of bioethics with other professional jurisdictions in the National Bioethics Advisory Commission's deliberations on stem cell research. Although bioethics has become a fixture in federal policy, particularly in the executive branch, bioethics faces formidable challenges in carving out a strong jurisdiction there, just as Evans (2002) found in his examination of the negotiation of the bioethics jurisdiction in public-policy debates on human genetic engineering, as reviewed previously in this chapter.

Bioethics Outreach to the Community

At the local, community level (rather than the federal level) bioethics has the potential to make an impact on issues by engaging with local organizations and social networks. Unfortunately, the Bioethics Center at Letters lost staff funding for outreach activities, as a result of downsizing efforts by the university medical system. The Jenkins Report reflected the disappointment of Center personnel about the Center's loss of the outreach function:

> In several meetings a variety of people expressed concerns about the outreach program and lack of community activities. Many felt the loss of the outreach program was a major blow to the [Center], and that it would be very difficult, although not impossible, to rebuild the valuable connections that had been fostered. They see community outreach as an important part of the [Center's] mission, and feel strongly that it should be restored if the [Center] is to maintain a regional presence in clinical and community settings.

After the completion of the Jenkins Report, new funding was located for outreach staff, and both new and restored activities were pursued.

In its first four years, the Center's annual reports emphasized the role of outreach in the Center's mission and in supporting of the Center. The first, second, and fourth annual reports all repeated the following statement "The [Center's] high visibility and numerous outreach programs make fundraising through private and foundation sources the most likely strategy for obtaining financial resources to support [Center] work." However, as the second and fourth annual reports both went on to note, "There remain serious failures on the part of the medical center and university to capitalize properly on the fund-raising opportunities that the [Center] presents." The fourth annual report described the repercussions of funding limitations for outreach activities:

> At current funding levels, [the Center] cannot maintain a high level of [outreach] support to the [Letters health system]…In addition, [the Center] must limit its conferences to only those underwritten entirely by corporate or foundation support…Finally, we need a way to create and market curricula for educating professionals and the public beyond conferencing. We want to reach out through standardized curricula in the areas of accreditation, ethics consultation, high school bioethics, and site-visit evaluations.

Restriction of outreach funding clearly limits the venues and audiences that can be targeted, and almost certainly the topics that are addressed, at conferences fully underwritten by outside organizations.

The scope of outreach recognized by the Center, and that could be pursued by the Center with its limited resources, is an important indicator of the Center's claimed jurisdiction. The preceding annual report statement indicates an aspiration to expand outreach activities in various ways, but is largely focused on a clinical audience. When outreach activities were renewed, subsequent to the release of the Jenkins Report, several outreach goals were enumerated, also largely focusing on a privileged, clinical audience. These goals included support for the health system's ethics committees; research on the function and efficacy of ethics committees; ethics application in venues like health care institutions, nursing homes, and consumer groups; and bioethics conferences organized by the Center.

One of my respondents commented that faculty "go out of their way" to give talks at local high schools, churches, and other community forums; they "try to have a local focus." However, she also noted that a sizeable proportion of the community in the area served by Letters were disenfranchised, and that "there's not a great relationship" between that community and the university, even though the university medical center's patient population comes from that community. She pointed out that the way participants are recruited for studies "raises some concern about whether you're using the population here like tissues and throwing them away... Lack of insurance brings up issues about patient injustice, consent forms; some of these people can't read, there are 12-year-olds with children."

However, there is little sign of outreach to these populations by the Center. One of my student respondents remarked that "universities always want to demonstrate their commitment to the community," but she was unsure how much university-wide outreach commitment there was at Letters. She was aware that a group of students ran a free clinic in the area, but said she was not familiar with the Center's community activities. She suggested it would be good if the Center operated an ethics hotline for the community.

Universities operate in networks of privilege, to some degree or another, which is often reflected in the type of outreach activities faculty engage in. Opportunities to dovetail outreach and fundraising activities have special appeal for academic units as they work to stabilize their operation, particularly for less stable, nondepartmental entities, and especially in lean fiscal times. As bioethicists seek to make permanent homes for themselves in universities, both for the sakes of their own livelihoods and the social values they hope to pursue, structural pressures will direct them toward outreach activities that facilitate institutional legitimacy and build social networks that will contribute toward organizational and fiscal stability.

CONCLUSION

This chapter examined the identity of bioethics and bioethicists, as perceived by faculty and students at the Center, and described the relationship of the jurisdiction of the Center to other jurisdictions at Letters University, as observed during my site visit. Overall, the influence of academic medicine on bioethics, and a motivation to forge relationships with potential sponsors and other sources of prestige and funding, appear to be the factors driving the construction of the bioethics jurisdiction at Letters, which serves in a weak advisory capacity to the biomedical jurisdiction. While health care professionals seem grateful for bioethics input and guidance for specific cases and projects, the Center's lack of engagement with the university medical center at the level of institutional policy suggests that bioethics at Letters stops short of systemic, structural critique of biomedicine, and remains "largely toothless."

Although the goals of bioethics were reported to be facilitation of dialogue about ethical issues in health care and biomedical technology, there remains work to be done by the Center in better facilitating dialogue on biomedical quandaries with stakeholders and disciplines (including nursing, IRBs in competitive research environments, academic philosophy, the social sciences, and marginalized groups in the surrounding off-campus community) which hold different perspectives and goals than bioethics. Bioethics at Letters University faces several obstacles in communicating with other academic subject areas and stakeholders both on and off campus, given the internal ideological tensions and fiscal vulnerability of the field.

Firstly, bioethicists are caught in the tension between the ideologies of social trustee professionalism and expert professionalism, the dominant ideology identified by Brint (1994). Pursuit of the ideology of public service requires bioethics to engage with other professions that are strongly influenced by the ideology of expertise. However, reluctance to clearly define the particular expertise of bioethics, and perceived weaknesses of bioethicists in key expertise areas including philosophy and the social sciences, make it difficult for bioethicists to be taken seriously and treated as intellectual equals by other professional experts.

Secondly, bioethics faces several structural challenges that place it in a myopic and reactive position with regards to the topics it addresses, rather than a farsighted and proactive position. Operating from the home base of academic medical centers makes bioethics fiscally vulnerable, and encourages bioethicists to be constantly vigilant for new funding opportunities.

Subsequently, they address the issues about which public and private patrons are sufficiently concerned to invest funds in. Close association with the mass media also presses bioethicists to analyze current social crises in a reactive, sensational fashion, rather than reflecting on, defining, and tackling typically unglamorous but chronic and consequential problems consciously selected by the field itself. Susceptibility to "biotechnofetishism" and a lack of demographic diversity among bioethicists also shape the set of topics on which the field focuses its attention.

Thirdly, bioethicists' skepticism of moral authority further mystifies other stakeholders regarding exactly what bioethics has to contribute to resolving ethical conflicts in biomedicine. Reluctance to don the mantle of moral authority is understandable given the roots of bioethics in championing patient autonomy and questioning the paternalism of the medical profession. However, skepticism of moral stances does not immunize bioethics against bias in the direction of particular societal viewpoints.

Rayna Rapp (1999) provided an instructive account on professional moral authority in her study of genetic counseling. The need to distance genetic counseling from the eugenics movement of the early 20th century led the field to embrace value-neutral, nondirective counseling. Rapp explained, however, that rejection of the ideology of eugenics simply resulted in the embedding of other values in genetic counseling. The focus of counselors (primarily females from white, upper-middle-class backgrounds) on the individual or couple as the presumed decision-making unit renders the role of social groups invisible in the counseling process, in spite of their importance in several ethnic populations and religious groups.

Genetic counseling also fails to openly acknowledge that the technology upon which it is based is not value neutral, but rather was developed to identify, and ultimately eliminate diseased fetuses. Furthermore, the ideal of nondirective counseling is not fully achieved; Rapp describes counselors challenging clients' rejection of prenatal testing for reasons such as fear of needles, and discusses allegations of counselors' activism against the use of prenatal testing for sex selection of fetuses. Like genetic counselors, bioethicists generally fail to recognize that the biotechnology which beguiles them is not inherently value neutral, and while ethics consultants may be averse to declaring moral stances at the bedside, bioethicists more than rarely express their personal positions on bioethical issues in the classroom and in statements to journalists.

Overall, bioethics appears to be quite similar to the fields of higher education studies and science and technology studies. These three academic fields were all created to manage a particular set of social problems associated with specific institutional settings and practices, and may never become fully institutionalized as common departments in universities. As a result of drawing on unlimited multidisciplinary tools to address practical problems, it is difficult for these fields to define an emblematic professional expertise with which to demarcate a freestanding academic jurisdiction. Bioethics may have some staying power by virtue of the outlook for the continued growth of biomedicine in coming decades, but the institutional and professional outlook of bioethics is unclear. Abbott's theory of the system of professions has limited ability to explain this Peter Pan occupational status, wherein an emergent group of expert professionals persist in a holding pattern that stops short of fully institutionalized professionalization (Abbott, 1988).

Undoubtedly, the ambiguous and problematic identity of bioethics as a field is not helpful in forging relationships with established disciplines outside the medical school or with stakeholders not immersed in academic culture, the conversational nature of the espoused goals of bioethics notwithstanding. For aspiring emcees in the ivory tower of babel, pointed dialogue within the field aimed at bridging gaps among the knowledge and skill bases of bioethics would likely also facilitate bridge building to neighboring expertise jurisdictions both on and off campus, each of which speaks in its own specialized language. Chapter 4 will examine the identity of bioethics students at the graduate level, and the logic of the master's programs in which they are enrolling.

ENDNOTES

1. At the time of my site visit, the Center was only accepting enrolled MBE students in its classes. Some of these students were enrolled in dual degree programs, such as MD/MBE or Nursing PhD/MBE.
2. The Task Force was convened by the Society for Health and Human Values and the Society for Bioethics Consultation, two of the three professional bioethics associations that merged into ASBH. The Task Force's 21 members included several prominent bioethics scholars as well as representation from the Joint Commission on Accreditation of Healthcare Organizations, the American Medical Association, the Society for Healthcare Consumer Advocacy of the American Hospital Association, the Department of Veterans Affairs, the Association of Professional Chaplains, and the American Association of Critical-Care Nurses. The Task Force took two years to develop the report, which was supported by funding from the Greenwall Foundation and 40 other organizations, including academic institutions and professional associations. Thirty-nine organizations submitted education and training materials for use by the Task Force. Over 1,400 copies of a discussion draft of the report were distributed to the bioethics community, and feedback was incorporated into a revised draft, which was reviewed and revised twice more by the Task Force before it approved the final draft that was adopted by ASBH.
3. Maryland and New Jersey have statutory requirements for HECs. See Hoffman (1991) and Scheirton and Kissell (2001).
4. Notably, attempts to document the impact of ethics education for improved moral decision making have failed to establish any benefit (see Bardon, 2004). For sociological critiques of HECs, see Bosk and Frader (1998) and Wolpe (2000). For a critique of the use of consensus in moral decision making, see Moreno (1995).
5. Interestingly, the Greenwall funding guidelines statement was recently revised to state that "The Foundation is especially interested in the work of junior investigators *and pilot projects that may lead to NIH support...*" This statement was retrieved January 22, 2008, from http://greenwall. org/guidebio.htm. The 2005 funding guidelines statements were retrieved October 10, 2005, from http://www.greenwall.org/exguide.html. Greenwall bioethics funding information was retrieved January 22, 2008, from http://greenwall.org/grantsSummary.php.
6. See Marker (2001).
7. Information retrieved January 22, 2008, from http://www.ncsl.org/programs/ health/genetics/rt-shcl.htm.

CHAPTER 4

MASTER'S DEGREE PROGRAMS
IN BIOETHICS:
A GROWING FRANCHISE

The recent growth and popularity of master's degree programs is not unique to bioethics, but rather mirrors a larger trend in graduate education in general. As of autumn 2007, there were nearly twice as many students earning master's degrees than there were in 1980, and since 1970 the number of master's students has grown at nearly twice the rate of bachelors and doctoral students (Fairfield, 2007). Fueling or perhaps fueled by the demand, the variety of master's program offerings has expanded as well.

'Boutique' master's programs—customized, interdisciplinary programs in trendy topic areas and often geared toward building practical skills for the job market—are being offered in specialty subject areas such as environmental science journalism (at Columbia University), literary biography (at New York University), and general humanities (at

New York University, University of Chicago, and Stanford University) (Schneider, 1999, p. A12).

However, the value of these specialty master's degrees is contested. Proponents argue that the programs are career driven, and can lead to higher paying employment with faster advancement than a BA alone, or may serve as a springboard to a PhD, offering "a kind of pre-professional initiation into academic discourse" (Schneider, 1999, p. A12). Critics argue that departments are simply exploiting students for tuition dollars, in order to offset financial losses from scaled-down PhD programs.

Indeed, Slaughter and Rhoades (2004) argued that the development of new "professional" master's programs represents a concerted effort by academic departments in U.S. universities to generate new revenue streams through such educational programs, but in a way that addresses the employment markets of the new economy. This effort is often tied to competition for internal resources, as some university administrators have begun linking student credit-hour production to allocation of resources to academic departments.

These master's programs, serving a continuing education function, are often relatively inexpensive for sponsoring departments because of the frequent elimination of the faculty-intensive thesis requirement. Furthermore, students in such programs require little research support, and often pay more tuition than doctoral students, who are more often subsidized by fellowships and assistantships in research or teaching. Slaughter and Rhoades (2004) explained, noting that business may sometimes make curriculum recommendations and provide monetary support:

> The development of new master's degrees is a dramatic break from the past and reflects a significant reorientation at the graduate level to the external employment market and to revenue generation. Part of the strategy is to forge closer ties to business. (p. 191)

These new master's programs are not geared toward training students for new employment, but targeted at students who are already employed and are aiming to upgrade or augment their skills.

Does the growing legion of master's degree programs in bioethics merely reflect the alleged trend toward "boutique" master's degrees, or does it address an actual workforce training need that students seek to fulfill? What, in fact, are students seeking from MBE programs, and what are they actually receiving? What is the value of MBE programs, for students, and for the future of the field? To address these questions, I shall first provide some context by examining faculty and student commentaries on the future field of bioethics, published in a 2002 issue of the *American Journal of Bioethics*. I shall then present findings from my case study of the Bioethics Center at Letters University, focusing on interviews with students. I was not able to collect data on tuition revenues generated by the MBE program at Letters; my analysis focuses instead on the perspective of students.

FACULTY AND STUDENT PERSPECTIVES
IN THE *AMERICAN JOURNAL OF BIOETHICS*

Recent commentaries from bioethics faculty members and students in a 2002 special issue of the *American Journal of Bioethics* (AJOB) provided insider perspectives on the functions and results of bioethics master's programs and the future of the field. like other master's programs, graduate bioethics programs, if they are primarily tuition driven, run the risk of enrolling too many students, spreading faculty too thin, and paying inadequate attention to graduates' career prospects (Kuczewski & Parsi, 2002). David Magnus, previously the director of a master's program in bioethics at the University of Pennsylvania, acknowledged the general benefits that master's students bring to bioethics centers, including being a source of revenue and research assistants for faculty. Another benefit, of particular relevance to emerging academic specialties like bioethics, is that of institutional legitimacy. Magnus (2002) quoted an unnamed colleague: "nothing makes you real in a university like a graduate degree" (p. 10). In the field of bioethics, master's programs are arguably preparing the way—and perhaps creating demand—for a growing number of new bioethics PhD programs. Charles Bosk (2002) argued that the

PhD "is nothing less than a legitimating chip, anted up in the bargaining for institutional resources and prestige" that consequently "create[s] competitive pressures on peer programs to do likewise" (p. 22).

Who is enrolling in master's programs in bioethics, and why? Unfortunately, detailed information on the demographics of bioethics students is lacking. Some students are recent BA recipients, viewing the master's in bioethics as a stepping-stone into a PhD program, or as a tool for deciding between graduate and professional schools. However, many other students pursuing master's programs in bioethics are midcareer professionals or dual-degree seekers, looking to specialize or enhance their current or imminent vocations, and not to switch careers (Kuczewski & Parsi, 2002; Magnus, 2002). Many bioethics centers explicitly state that their programs are not designed as a means to pursuing a career (Magnus, 2002; Russo, 1999). Chaitin (2002), one of the student commentators in AJOB, noted that even her PhD program in health care ethics at Duquesne University "cautions students that a degree earned through its program will not be a guarantee of a future in bioethics" (p. w2). While some graduate programs publicize such caveats, the 2001 American Society for Bioethics and Humanities (ASBH) graduate program survey found that a "significant minority" of master's program do claim to train individuals for full-time bioethics-related work (Ausilio & Rothenberg, 2002, p. 7).

Whether or not bioethics master's programs are aimed at job training, AJOB commentators see implications of graduate education for the bioethics-related job market. Russo (1999) contended that clinical ethics consultation has not generated a vigorous job market, and hospitals may instead be meeting accreditation requirements via the use of unpaid ethics committees. Magnus (2002) speculated that it may not be necessary to employ separate clinical ethicists on staff at all. Instead, it may be more effective "to spread clinical staff with some bioethics education throughout the institution, ideally as part of every team" (p. 12). Alternately, holders of PhDs and JDs who are currently employed in clinical ethics positions may be displaced in the near future by less expensive master's recipients (Magnus, 2002).

Steady growth in the population of bioethics degree holders will create a buyer's market in bioethics jobs, predicts Charles Bosk (2002), a state of affairs, he argued, that does not encourage independent, critical thought among practitioners. "Typically, the master's degree is not associated with independent practice," he explains, but rather with "administrative positions in professional bureaucracies," as in the case of social work (p. 22). Although steadily growing in number, it is not clear the master's in bioethics (MBE) programs will have a positive impact on the future of this relatively young and undeveloped professional field.

From their observations as bioethics professors, Kuczewski and Parsi (2002) argued that the motivations of MBE students "generally sum up to the broader aim of being a member of the bioethics community" (p. 14). In fact, the authors consider it to be the "only one [student motive] that can account for the rapid growth of terminal M.A. programs [in bioethics]" (p. 14). They acknowledge intellectual curiosity and the desire for professional credibility as other student motivators. However, students are not merely seeking access to bioethics knowledge, which is increasingly accessible. Rather, they desire sympathetic interlocutors for ongoing support and mentoring. Kuczewski and Parsi (2002) explained,

> Ethical issues are often complex and *implementation of new approaches tends to be difficult requiring long-term strategies and persistence* [italics added]. For instance, seemingly simple issues of informed consent and quality end-of-life decision making *have proven recalcitrant and are continuously inviting renewed intellectual efforts. Maintaining morale and the will to address such matters* [italics added] requires the support of a professionally trained community. (p. 15)

This account conveys not only the frustration regularly faced by professionals in the clinic, but also suggests that bioethicists serve both as confidants and as consultants, sustaining morale as well as providing advice for recalcitrant problems.

Harrison (2002) adopted a perspective on the functions of bioethics that is more competency based. She lists particular skill areas she finds necessary for success in bioethics, including conflict resolution, critical evaluation of evidence, group processes and dynamics, research methodology, and policy analysis. Harrison appears to question whether these skills are being taught effectively, noting that in her own graduate work, such skill acquisition was "coincidental" to the completion of degree requirements, and mentoring was "haphazard" (p. 20).

In contrast, Kuczewski and Parsi (2002), respective directors of the bioethics center and online master's program at Loyola University, Chicago, contended that MBE programs are shaped by student aims. Reflecting on the history of MBEs, they recount "the main story many of us have been telling," in which the first programs were launched, in response to increased attention to medical ethics issues, as tracks in traditional humanities departments (e.g., philosophy) (p. 13). The enrollees were largely clinicians, presumed by program designers to desire a foundation in traditional aspects of a discipline such as philosophy, although some students sought an applied ethics degree to supplement doctoral work in other fields. The programs consequently tended to consist largely in typical disciplinary graduate coursework augmented by a few specialized courses developed for the track.

However, this approach was somewhat misguided, because the clinician students did not perform as well as the traditional graduate students did with regard to coursework, and not surprisingly had a more practical than scholarly orientation. One can readily imagine pragmatic students with MDs becoming quite exasperated from studying the decontextualized analytic intricacies of moral theories in a rigorous graduate course geared toward PhD students in philosophy. Kuczewski and Parsi (2002) argued that the trend in the 1990s, away from disciplinary tracks and toward the founding of new "genuinely interdisciplinary" bioethics programs, has come about because "no one discipline made a convincing case that bioethics is simply a subset of it [that one discipline], [thus] more diversified offerings are sought" (p. 14).[1] This trend suggests to the authors "that the concerns of persons enrolling are fairly practical in nature and likely to be ongoing" (p. 13).

Student Perspectives

In addition to providing space for faculty commentary on the implications of the 2001 ASBH North American Graduate Bioethics and Medical Humanities Training Program Survey (ASBH, 2001), AJOB also presented feedback from current and former graduate students in bioethics/medical humanities programs. To guide the responses, the AJOB editors asked each student commentator to consider three questions:

- Why did you enroll in a graduate bioethics/medical humanities program?
- How well did your graduate bioethics/medical humanities program meet your needs?
- What do you think graduate programs like these mean for the future of the field?

In the 24 student responses, I examined themes related to motives for program enrollment, program features of salience to students, and satisfaction with program experiences. Respondents were diverse in many respects, representing different student backgrounds, institutions, disciplinary orientations, and credentials. After reviewing themes in student responses, I shall discuss the correspondence between student and faculty perspectives and claims.

Student Motives

Reasons given for pursuing graduate credentials in bioethics fall into the categories of need relating to existing jobs, love of bioethics, and related career aspirations. Butkus and McCarthy (2002), students in the PhD track of the Healthcare Ethics program at Duquesne University, were both motivated by the need to learn and integrate both the abstract theoretical foundations and the tangible clinical experience they required for their work. Both students chose Duquesne because they felt it would provide them with the desired groundwork for ethics consultation and teaching.

Other commentators aspired to becoming well-rounded clinicians, and also to influencing their colleagues, with the help of a bioethics education. Bevins (2002), a medical student, explained that obtaining a foundation in the medical humanities was essential to becoming the "kind of doctor" and the "kind of person" he wanted to be (p. w1). Unsurprisingly, he also notes that his program's location at an academic medical center (University of Texas Medical Branch) provided a rich environment for pursuing diverse interests and goals.

Nathanson (2002), a psychiatric resident, stated that "the current level of bioethics education at most medical schools is severely lacking (p. w1). He felt that his graduate studies in bioethics (a master's in public health from Boston University, with a concentration in health law and medical ethics) gave him specialized knowledge and skills in the analysis and resolution of moral issues that will be invaluable to his future practice.

Egan (2002), an internal medicine resident, identified bioethics as her medical "specialty" and plans to maintain a general clinical practice as well. Like Nathanson, she found that medical humanities training for medical students is wanting, and decries most physicians' deprecating attitude toward ethics training and their habitually unethical behavior in clinical practice. Egan (2002) argued that "residents need to be taught that passive ethics education does not prepare them for the ethical challenges they will regularly face in practice" and that "clinicians not only need more ethics training exposure but also a reformulation of the role of ethics in daily practice" (p. 27).

Other student commentators sought skills in research or clinical ethics. Veikher, an international student from Russia, cites an urgent need in Russia for experts in international research ethics. She was one of four students chosen by the Fogarty International Center to receive fellowships to study bioethics at Case Western Reserve University during the academic year 2001–2002. According to Veikher, the Fogarty Center selected Case Western Reserve University because its MA program provides a general theoretical and practical introduction to the field of bioethics, which is most useful to international students. She particularly values the opportunities to learn consensus-building skills and to acquire practical experience through clinical rotations (Veikher, 2002).

Building clinical ethics skills was an important motive for other commentators. Chaitin (2002) is an intensive care social worker, hospital ethics director, and medical school faculty member. Although the combination of her MA and PhD education in bioethics enriched her practice by providing her with a "stronger holistic knowledge base," (p. w2) she is not satisfied with the training in ethics consultation provided by her PhD program. Gurin (2002), an MD who increasingly found herself playing the role of ethicist, sought graduate bioethics education to acquire essential tools for addressing clinical ethics issues, something which would benefit not only her own work, but also her entire hospital community. McCruden (2002) began his career as a hospital chaplain and later came to serve on ethics committees and as a health system administrator. He desired to increase his familiarity with clinical ethics issues and theoretical approaches in order to better perform his roles, and particularly to apprehend patient rights and organizational ethics "in order to remain a competent resource" in his workplace (p. 25).

Beyond immediate job-related needs, many student commentators also cite love of bioethics and clear desire to work in that field as a motivations for pursuing graduate training in bioethics. Thaddeus Pope (2002) stated simply that he enrolled in Georgetown University's JD-PhD joint-degree program "to become a bioethicist" (p. 37). Strauss (2002), another graduate of Georgetown, reflected, "the field of bioethics satisfied my passion for the contemplative life, as well as my need to perform a public service" (p. w1). Schonfeld (2002) became "totally enamored" with bioethics as an undergraduate "languishing somewhat aimlessly with a religious-studies major" (p. 32). He subsequently pursued a PhD, and became a professor of medical humanities. Solomon (2002) explained that while she had an interest in bioethics, she also needed a practical reason to study bioethics. As a secondary school science teacher, she began to find that more and more bioethics-related topics began to present themselves in her classroom, and realized that she wanted to teach bioethics.

Several other student commentators might be categorized as eclectic wandering scholars—reflective persons with diverse interests who eventually found a home in bioethics. Sisti (2002) explained there was no single reason why he was drawn to bioethics, but "rather an amalgam of interests and goals somehow coalesced" and led him to what was, for him, "the perfect graduate program" (p. 28). For Voss (2002), who found the way down life's path often "tortuous and confusing," health care ethics is "a field that invites those with diverse backgrounds to help those dealing with serious bioethical dilemmas" (p. w1), and one where he could use both his veterinary and theological training. McGuire (2002) went from being an undergraduate psychology major to studying shamans in the Amazon rain forest before she pursued a JD-PhD joint-degree program in hopes of achieving "a unique perspective on the legal and ethical issues involved in western medicine" (p. w1). Suziedelis (2002) viewed her admittedly eclectic academic background in education, political science, and theology as "the perfect preparation" for doctoral work in health care ethics (p. w1).

Other student commentators state that their career aspirations led them directly or indirectly to seek graduate-level bioethics training. Isinger (2002) developed an interest in bioethics as an undergraduate, and chose to pursue a PhD because it "opens more opportunities for employment and advancement," thinking that to succeed in a bioethics career she would need a "professional appointment, and academia was the logical choice" (p. w1). Zilberberg (2002) enrolled in a philosophy PhD program with a view to becoming a philosophy professor, and, upon taking a bioethics course, realized that she really wanted to pursue the field. McCarthy (2002) knew as an undergraduate that what she "loved more than medicine was the art of teaching about medicine," and soon found herself searching for a graduate program in healthcare ethics to pursue a teaching career in medicine (p. w2).

Key Program Characteristics

Student commentators cite a variety of program characteristics as important to their educational experience. Salient features included interdisciplinary characteristics, program flexibility, a balanced curriculum, opportunities for clinical practica, effective mentoring, and the possibility of gaining particular skills necessary for work in bioethics.

Some commentators highlight the role that a sense of community played in their graduate training. Butkus and McCarthy (2002) noted that the students in their program "have a remarkable sense of community and a desire to assist one another," which contributes to a "positive, committed environment" in the program (p. w2). Bevins (2002) described the shared mindset of participants in his graduate program at the Institute for the Medical Humanities at the University of Texas Medical Branch, which includes, among other things, "a belief in the importance of language as a means of individual expression and social change," "a belief in an individual's ability to develop himself or herself," and "a commitment to affecting change through public discourse" (p. w1).

Several student observers comment positively about the disciplinary diversity of the training and the student body in their programs. Butkus and McCarthy (2002) viewed student diversity positively, stating that the graduate program they both attended enabled them to build "an excellent knowledge base in law, philosophy, psychology, clinical medicine, and theology," and "to develop a network of professional contacts in all areas of medicine and academia" (p. w2). This network is not necessarily due to student diversity, but rather to the array of professionals with whom they worked during their program experiences.

McCarthy valued the opportunity to receive a "strong theoretical grounding in a variety of sources—religious, cultural, and philosophical," and to partake of "a wide range of clinical experiences" (Butkus & McCarthy, 2002, p. w2). Zoubul (2002) found student diversity was beneficial to her graduate experience, explaining that "by combining professional and traditional students from a variety of academic disciplines," the diversity of her graduate class "allowed for insight from many perspectives" (p. w1).

Isinger (2002) found both the international and professional diversity among her Duquesne classmates to be valuable because it generated enlightening commentary stemming from different theoretical frameworks (p. w1). Suziedelis (2002) commented that the interdisciplinary nature and flexibility of her program

> makes it ideal not only for students with diverse academic backgrounds, but also for those such as I, who bring to it a wide range of life and educational experiences that we seek to pull together into new, integrated, midlife careers. (p. w1)

Program flexibility was an attribute that figured significantly into several students' assessments of their experiences of bioethics graduate training. McCruden (2002), a hospital administrator, found that the online graduate bioethics program offered by Loyola University, Chicago, alleviates the constraints of time and travel presented by typical continuing education programs in ethics, meeting his needs for a highly accessible and flexible training program. He speculated that many professionals from various backgrounds are interested in expanding their knowledge of ethics, but the "lack of a program close to home that fits into a busy schedule" is a significant obstacle (p. 26).

In addition to convenience, McCruden (2002) argued that with a good instructor, Web-based education can be very student centered, observing that Loyola "has a reputation for having the most responsive and innovative faculty" (p. 26). Chaitin (2002) also mentioned that her doctoral program at Duquesne University "seems to be designed for the working adult," offering night classes and online services (p. w1). McGuire (2002) asserted that "if graduate programs in bioethics and the medical humanities are to survive...they must adapt to the changing needs of their students;" she noted that she took advantage of her program's flexibility to take classes in various departments, to arrange an extended clinical practicum, and to conduct her own research (p. w2). Her fellow University of Texas Medical Branch (UTMB) student, Bevins (2002), agreed that the medical humanities curriculum offered there is "flexible enough to accommodate people from all sorts of backgrounds with all sorts of interests and goals," whether they want a broad education or a more focused one (p. w1). Acknowledging that no one program can meet all needs perfectly, Isinger (2002) recommended that bioethics graduate programs provide some flexibility in what courses are required, to assist students in filling their particular knowledge gaps (p. w2).

Student commentators also spoke of the importance of a balanced curriculum in graduate bioethics training, with some alluding to what Butkus and McCarthy (2002) called "the disparity between clinical and classroom ethics" (p. w1). Veikher (2002), an international student, noted that in Case Western Reserve University's program the core courses provided both theoretical knowledge and practical experience in the curriculum. Butkus and McCarthy (2002) testified that they received a thorough education in an array of relevant disciplines, as well as extensive applied clinical experience; their program not only trains students to competently examine the array of ethical, legal, and social challenges raised by modern medicine, but also accommodates the wide array of clinical and nonclinical backgrounds which students bring to the program. McGuire (2002) echoed that graduate bioethics programs must, among other things, take responsibility for "integrating theoretical studies with clinical experience and research" (p. w2). Zoubul (2002) noted that at Case Western Reserve University, curricular inclusion of "academic coursework and clinical observation allowed us to see bioethics as it is done in two very different settings, the classroom and the clinic" (p. w1).

Some students judged their programs to be lacking in a balanced curriculum. As an experienced clinical consultant, Chaitin (2002) found her master's education in bioethics to have been lacking in adequate attention to the clinical environment and its legal context, and deemed the clinical experience offered by her doctoral program to have been poorly designed and supervised. Schonfeld (2002) commented that "in many programs there is often a dissonance between classroom education and practical application" and suggested that including more team or committee experience in the clinical ethics training of graduate bioethics programs would help students to better understand the role of the ethicist in the collaborative clinical setting (p. 33).

Several students specifically emphasized the importance of clinical practica in their training. Schonfeld (2002), now a medical humanities professor at the University of Nebraska Medical Center, noted that "part of the goal of the [graduate] practicum was to develop my sense of how a clinic worked, but another part was also to understand the role of the ethicist in the clinic" (p. 33). White (2002), a former student of UTMB, found that "total immersion in the culture of medicine was inescapable and a very important part of my education" (p. 34). However, when she began teaching medical students, she realized that "a primary weakness of my graduate training was that I had not spent enough time with the kinds of practical clinical information my students demanded" (p. 34).

Mentoring and role modeling were also repeatedly mentioned as crucial features of good graduate bioethics training, particularly with regard to the clinical aspects of training. Sisti (2002) credited the effective mentoring he received at the University of Pennsylvania bioethics program with helping him parlay a clinical ethics internship into a "more-or-less full-time clinical bioethics role in a Philadelphia health system" (p. 29). He also speculated that neophytes in bioethics are likely to become discouraged unless they receive the opportunity to develop "prudent reasoning through solid mentoring" (p. 29).

Schonfeld (2002) argued that role modeling and mentoring are an important part of professionalization, teaching students the roles and skills needed to become an ethicist, beyond what can be taught in the classroom. He commented that he would have liked the opportunity to shadow ethicists in the clinical setting in the same way he and his fellow students shadowed health care professionals, but also noted that this would have required particular effort, given that the faculty ethicists in his program are "primarily academic philosophy professors and do not spend much time in the clinic" (p. 33). Strauss (2002), voiced an opinion similar to that of Schonfeld, stated, "I do not think proper professional development is possible without good mentorship," acknowledging her own mentors for modeling "ideal professionalism" and providing her with opportunities (p. w2).

Chaitin (2002) reflected that "the provision of a mentor [during clinical training] was invaluable to me for many reasons, not the least of which was a need to learn how to view what I was observing with different eyes;" her mentor taught her "to see a problem clearly through a prism rather than a magnifying glass" (p. w1). Isinger (2002) similarly advised bioethics graduate programs at all institutions that mentoring for the first student practicum would be helpful. Zilberberg (2002) concluded, "the aspect of the program that has most helped me become a bioethicist and build a career is the mentoring and advice of helpful faculty members" (p. w1).

Several student commentators emphasized the importance of particular skills or tools they felt they had gained during their graduate bioethics training. Interpersonal capacities including communication skills, consensus building, and tolerance were cited by some students. Schonfeld (2002) really valued learning to make presentations to clinical audiences and subsequently feeling comfortable speaking to health care professionals. Veikher (2002) and Chaitin (2002) both acknowledged the value of exposure to conflicting opinions, different religious perspectives, and various theoretical approaches, which their respective programs afforded them; they thereby learned about themselves, about others, and about tolerance and achieving consensus. Other students prized the analytical skills they developed in their programs, including the ability to navigate complex ethical issues and the theoretical and topical complexities of bioethics, and to analyze and resolve ethical problems.

Program Satisfaction

Regarding the level of satisfaction with their experiences of graduate education in bioethics, several student commentators made positive remarks, while several others expressed concerns about their future career prospects, and about the role of graduate programs in the future of the bioethics.

Satisfied program graduates feel that their experiences not only provided contributions to their individual professional growth, but also to a larger community. Pope (2002) felt that his program exceeded his expectations, stating

> I went to Georgetown to become a bioethicist. Once there, I discovered that the role of the bioethicist is no more unitary than that of the lawyer or the doctor. I soon decided what **sort** [emphasis original] of bioethicist I wanted to become, and Georgetown prepared me well for that role. (p. 37)

Campbell (2002) commented that her program exceeded her hopes and that "while I don't pretend to be an expert in the field...I have the tools that will enable me to make a meaningful contribution to the field" (p. w1). Gurin (2002) stated, "the entire community [at her hospital workplace] is benefiting from my education;" she serves as the only ethicist on her hospital's institutional review board, and also as the only physician on its Ethics Facilitation Service (p. 34).

Veikher (2002), who will take what she has learned about research ethics back to Russia to fulfill an urgent need, felt that her program was "an outstanding success" as an experiment in placing an international student in a bioethics program at an American university. Butkus and McCarthy (2002) wrote, "We are beginning what we envision to be long careers in bioethics, and we believe that our program has provided us with an excellent knowledge base" and "has allowed us to develop a network of professional contacts in all areas of medicine and academia... We could not have asked for more" (p. w2). Bevins (2002) similarly concluded that he could not have made a better choice than the UTMB program, and that he "will be a better doctor for it" (p. w1). Isinger (2002) likewise conceded that while no program can perfectly meet all needs, she would do it over again.

Commentators expressing career concerns spoke of the need for more jobs in the field, the value of certification in research methods, the role of graduate work as a career stepping-stone, and the role of graduate programs in the future of bioethics. Zilberberg (2002) felt there is a need to create or locate more jobs in bioethics, and Voss (2002) deemed that pursing an additional certificate in empirical ethics research methods would "greatly enhance my chances of securing a desirable academic position" after graduation (p. w1).

Several comments made concerned the role of graduate work as a career stepping-stone. Shivas (2002) argued that interdisciplinary coursework and other interdisciplinary experiences are " a real asset to those hoping to pursue a career in bioethics" (p. 24), but she also cautioned, "an interdisciplinary program on the PhD level may leave one without the necessary theoretical grounding through which those interdisciplinary skills can be utilized and allowed to flourish" (p. 25). Sisti (2002) warned, "young scholars…should not enter a bioethics program with the dream of coming out a card-carrying 'bioethicist'" (p. 29), but he conceded that "increasingly, jobs *are* [emphasis original] out there" (p. 29), and that "the doors have swung wide open for the professional domain expansion of bioethics and its practitioners" (p. 29).

Zoubul (2002) felt that her graduate education and work experience would put her in a better position to meet her professional goals. Chaitin (2002) believed that her education enriched her practice as a clinical ethics consultant, and that she was mistaken in her previous belief that what she had learned from her clinical work as a social worker was sufficient preparation to be a successful ethics consultant. Zilberberg (2002) reflected, "the aspect of the program that has most helped me become a bioethicist and build a career is the mentoring and advice of helpful faculty members" (p. w1). Nathanson (2002) thought that his program provided necessary skills in ethical analysis and problem solving that will be invaluable in his future practice, and that his research experience in health law was indispensable.

Two students commented on the relationship between graduate programs and the future of bioethics. Isinger (2002) believed that "'specialized' degrees in this field will help establish and perpetuate the profession" (p. w1). However, she also noted,

> Most professions, when trying to establish themselves and gain recognition, usually undergo a process of monopolization and regulation of skills and knowledge. However, the multi- and interdisciplinary nature of bioethics/medical humanities is what makes the profession unique, interesting, and challenging. I hope that those of us trained specifically as bioethicists never come to believe that it is only 'experts' who qualify to 'do' bioethics. (p. w1)

Zoubul (2002) affirmed that graduate programs play an important role in bioethics. "Although our professors spent a lot of time stressing the fact that we should not view the M.A. as a terminal degree," she observed, "I think that graduate programs like this are integral to the future of the field" and "can be a stepping-stone in building a career in bioethics" (p. w1–w2). For Zoubul (2002), the value of the MBE is clear: "the combination of my graduate education and work-related experience in bioethics has put me in a much better position to accomplish my professional goals than if I had not completed the program" (p. w2).

Summary of AJOB Student Commentaries

What do students want from MBE programs, and why? Are they getting it? The student commentators in AJOB came to bioethics with a diversity of backgrounds and goals, and entered graduate bioethics programs which had a variety of formats and orientations. Many have emerged or can be expected to soon emerge satisfied, on balance, with their educational experience, and most clearly articulate particular aspects of their training that they deem important. Several MBE students are midcareer professionals supplementing their skill sets, while others are embarking on careers as not only self-described, but also credentialed bioethicists. Of course, one would not expect disgruntled students and dropouts to write and submit accounts of their experiences (and even if the were to, perhaps such accounts would not be published), so one cannot conclude, on the basis of these commentaries, that most students are satisfied with or gain career benefits from their graduate education. Several commentators offered constructive criticisms of their respective programs and of the field of bioethics in general, but none were scathing. It is noteworthy that several students refer to themselves as bioethicists, can clearly identify job prospects in bioethics, and even entered programs explicitly to become bioethicists, in spite of acknowledged warnings that graduate programs are not geared toward helping one find a job placement as a bioethicist.

How do student responses align with faculty member commentaries in AJOB? The volume of student comments about the importance of mentoring in their graduate experiences is consistent with Kuczewski and Parsi's (2002) allegation that students primarily desire to become a part of the bioethics community and to receive support from it. However, students also spoke at length about the invaluable skills, knowledge, and firsthand experience that they desired from their graduate bioethics training, indicating that content as well as community is important. Students also related specific career needs and goals to their training; they sought mentoring and practical experience as well as an interdisciplinary perspective and knowledge of theory and methods, to enhance their current work in ethics teaching or consultation or to prepare them for careers in which bioethics will be either the primary focus or an enduring component.

The discourse published in AJOB on graduate bioethics training conditionally suggests that MBE programs provide valuable preparation to students, as well as tuition revenues and legitimacy to academic bioethics programs. I now turn to my case study findings to learn what value the MBE program at the Letters University Bioethics Center provides to its students and faculty members.

THE MASTER'S PROGRAM AT LETTERS UNIVERSITY

Planning documents for the Center's MBE program differentiate it from traditional master's degree programs. In a letter discussing the arrangements for the fledgling master's program, one administrator observed that the bioethics program "is distinguishable from an MA or MS degree *that is a stepping-stone towards a PhD* [italics added]," and consequently, that "the new Master's would not shape or influence future decisions about creating a PhD in Bioethics." An internal memo regarding the Center's marketing plan for the master's program expressed that the marketing goal was to double, if possible, the number of midcareer professionals applying to the program over the third and fourth years of the program.

The marketing materials for the MBE program explicitly indicate that applicants should be midcareer health professionals, or students already admitted in other graduate or professional degree programs at Letters. Prospective students with merely an intellectual interest in bioethics are also invited to apply, but only with the provision that they understand that the MBE, by itself, is inadequate for finding a job placement.

These observations indicate that the bioethics master's program was not conceptualized as a traditional academic program, but as a training program to augment the careers of established professionals.

In the marketing-plan memo, a staff member shared marketing ideas which focused on the local health care community. The resulting list of marketing strategies provides a sense of the vast network of the Center's potential constituents. On campus, these include schools, departments, centers, institutes, public-relations offices and ethics committees all over the university, clinical affiliates of the medical center, and continuing medical education representatives. They also include extramural institutes, hospitals, managed care companies, health care and ethics organizations in the region, professional health care conferences held in the vicinity, and local chapters of health professional associations and patient/consumer advocacy groups. Other constituents mentioned were regional health care publications and government offices such as the health departments. A related Center meeting agenda also cited pharmaceutical companies as a promotion target. Clearly, the Center saw myriad opportunities to connect with an extensive network of interests and organizations in its environs, and intended to tap into it.

According to the graduate program director for the Center, the master's program in bioethics was part of an initiative in the College of Arts & Sciences (CAS) to develop a set of professional master's programs in its Division of Continuing Studies (DCS). The dean of DCS, I was told, was an entrepreneur who wanted to expand offerings for a new market, midcareer professionals, as a way of bringing in revenues. At least one of these programs (as of October 2005, there were six such DCS master's programs) was highly successful and grew quite large. Concerns emerged about the quality of these programs, due, in part, to the fact that many courses were concurrently offered to graduate and undergraduate students, who attended the same class sessions (this was not the case for the MBE program). Furthermore, although DCS had expertise in dealing with adult students, the division was not well run from an administrative point of view, according to the MBE program director. Because key faculty from CAS were not available to take major roles in the new program, the original vision for the MBE to be a CAS program was not feasible. Most of the coursework was consequently offered out of the Center, not through CAS. At the time I visited the Center, the MBE program, and its revenues, were about to be moved from CAS to the medical school, and the Center had considerable control over the program.

One would expect that operating the MBE program entirely out of the Center might place a considerable work burden on Center faculty. The Jenkins Report, an independent evaluation of the Center, to which I referred in chapter 3, suggests that Center faculty were indeed spread thin. The report cites student concerns that the classes were too big and that faculty members were already too busy for the program to expand further. Many students also expressed the desire for faculty to provide them with more structure and direction, particularly in their clinical experiences. The report also states that some students received the impression that the program was "just added on" to the Center, and was not a priority. It would appear that perhaps the Center had an overly ambitious timetable for growing the MBE program, both in terms of its faculty-student ratio and the overall faculty workload.

Students at the Center

At the time I visited the Center, the graduate program director estimated that more than two-thirds of the MBE students were midcareer professionals, with the vast majority ranging in age from the late 20s to the mid-50s. The remainder of the student body consisted of dual-degree seekers (law or medicine in tandem with bioethics) and a few recent bachelor's graduates. The graduate program director also explained that while the largest fraction of students were physicians, "we put a premium on diversity of experience," as evidenced by the nurses, veterinarians, psychologists, lawyers, pharmaceutical industry professionals, consultants, clinical researchers in the Center's student body, not to mention the other professions represented. She estimated that 8% of MBE students had been African American, and that smaller percentages were Asian or Hispanic, with an approximately equal representation of the genders. By comparison, one third of the Center faculty members are female.

With the help of a Center staff member, who sent out sent out invitational e-mail messages via the Center's student listserv, I solicited volunteers from the Center's current student body and alumni for individual and group interviews. Only 11 students completed interviews, all of which were conducted one-on-one; 7 of the students were female. All of the students interviewed were white, and 2 were from abroad. They ranged from approximately 20 to 60 years of age. One program alumnus had earned an MBE while in medical school, whereas all the other interviewees were early or midcareer professionals. Two were program graduates, 1 of whom acquired a staff position at the Center after graduation; 3 were attending part-time over an extended term, 1 as a dual-degree seeker (MBE in bioethics, PhD in nursing); and the remaining 6 students were in their second year of the program. From the group of currently enrolled students, I interviewed 3 physicians, 2 nurses, 2 lawyers, 1 veterinarian, and 1 individual with a public-policy background. For the sake of anonymity, where pronouns are required for readability, I will refer to student respondents with feminine pronouns.

I conducted taped semistructured interviews with 10 of the students, most of which lasted less than an hour. One student declined to be taped. In general, the student responses I received were less rich than the perspectives shared by student commentators in AJOB, which I examined earlier in this chapter. Students' responses were grouped under five themes drawn from the interview questions: their perceptions of the Center, their motives for entering the program, their expectations of the program, their perceptions of the value of a master's degree in bioethics, and their perceptions about the goals of bioethics.

Perceptions of the Center
Student interviewees conveyed some shared views of the Center. Five described the student body as diverse in terms of their backgrounds and interests. Two described faculty as difficult to access, and 3 described faculty as biased in their views. Students were unclear as to the Center's role in the larger university or local community contexts. While 2 of the interviewees cited the Center as playing an active role in the

medical school, a third interviewee expressed surprise that the Center did not play a larger role on campus. A program alumnus who had become a Center staff member observed that while there were some collaborations with faculty in other parts of the university, the Center largely operated separately from the Letters University community. When probed, only 3 of the students interviewed commented on external factors or influences on the Center, and all 3 cited current events as a significant influence. Two students cited the need for funding as an important pressure, one of whom more specifically identified competition with other bioethics centers for money and recognition as a salient influence.

Student Motives

Students cited a variety of career goals and other motives for pursuing an MBE and choosing the program at Letters University. Several mentioned teaching, consulting, and public-policy work as career goals related to bioethics. Motives for pursuing an MBE included personal enrichment, curiosity, the acquisition of job-related expertise, and the legitimacy or marketability associated with holding the degree. Seven of the students interviewed cited proximity as a major reason for choosing the Letters University program, although several also cited the Center's reputation as an important draw. One student relocated specifically to pursue a dual degree at Letters University, and another dual-degree seeker commuted a long distance, feeling that Letters had more to offer than other universities in the region.

Student Expectations

Students commented on their expectations of and satisfaction with the program. While 3 students said they had no idea what to expect when they entered the program, others had explicit expectations. Three anticipated obtaining a good grasp of the foundations of bioethics, and 1 dual-degree seeker had hoped that the MBE program would dovetail better with her other program of study. One student thought that bioethics would have a broader perspective and more openness (addressing environmental concerns, for example) but found that bioethics as presented by the Center was "strictly medical ethics."

Students were mixed with regard to their satisfaction with the program. One student, a dual-degree seeker in nursing, was "really pleased" with the experience, 6 students were mostly pleased with what they encountered in the program, but felt there was room for improvement, and 4 expressed significant disappointment. Students' critiques related to the professional conduct of a few faculty members and to the curriculum. Two students felt the curriculum could be more balanced. One wanted more of a clinical focus, and felt that students received a biased presentation of the field. She also commented that "unsexy topics" were not well addressed. Another interviewee felt that there was too much philosophy, and wanted more theology in the program.

The Value of the MBE

I asked students to reflect on the value of the program, and on the expected utility of an MBE. The students felt they were gaining both epistemic and social capital from their program participation. For these students, epistemic capital consists of learning the fundamentals of bioethics— a working knowledge of the core topics and literature, the key issues, the philosophical underpinnings, and history of the field—and of developing skill in analyzing ethical issues, and dissecting and formulating ethical arguments. With respect to social capital, students cited the benefits of becoming integrated into the professional bioethics community. One student remarked that she could now "act like and talk to bioethicists," and 2 mentioned that participating in the MBE program boosted their confidence. Two students mentioned participation in the American Society for Bioethics and Humanities conference as a benefit, and a third student mentioned the value of acquiring mentors to keep in touch with.

When further probed about the utility of the degree program to them personally, with regard to their careers, and in terms of their futures, students talked about the value of the credential, and how it would distinguish them from other job seekers lacking an MBE. One student felt the credential would make her "marketable," and help with obtaining a teaching position in a medical college, particularly since few other nurses or clinicians have an MBE. Two other students mentioned the utility of the degree for obtaining positions teaching ethics. Another commented that the MBE will give her a "niche," and that it would "open doors" and "set her apart." One student noted that her training "opened up a new way to discuss things" in the career she is pursuing. Other students mentioned that the degree would bring them "cachet" or recognition.

I asked students whether they had participated in faculty research, or had worked with faculty in any other capacity. Four students expressed regret at not having worked with faculty, due to reasons such as distance from campus, the constraints of dual-degree programs, or otherwise not having enough time. Only 2 of the students I interviewed had worked with program faculty; another student had worked with a clinician not among the Center faculty, and 2 others indicated that they might pursue work with faculty members in the future.

The Goals of Bioethics

Students were asked for their views on the goals, or task, of bioethics, and on what progress had been made toward those goals. Overall, students' responses were consistent with the view that it is the job of bioethics to raise questions, "to enlighten," not to provide answers. As one student put it, the aim of bioethics is to "look at how we can make better decisions around issues affecting the human body and relationships with health care providers." Another student said that, in addition to questioning the technological imperative, and to emphasizing the importance of considering patients' interests, it was the job of bioethics to ask, "Whom does this [biomedical] technology serve?" and "What are our values?" Similarly, one student felt that bioethics is a means to addressing science's responsibility to society. Another commentator echoed that it was probably not the task of bioethics to serve as a "watchdog," but rather as "a buffer between what science and medicine can do versus [what science and medicine] should do."

A few students suggested slightly more active goals for the field of bioethics, such as improving policy and practice in areas like research ethics.

One student stated, "bioethics is a diplomacy tool; we're not just referees," and went on to explain that bioethics brings an analytic framework to discussions, that the role bioethics plays is akin to brokering negotiation—a bioethicist needs to translate between views, to be a diplomat, to bring stakeholders to a common consensus. Two dissenters argued that bioethics is ideology, with one summing up the field as "WASP ethics," and another arguing that bioethics is a "secular religion," providing counterargument to religious arguments on bioethical issues.

Students differed regarding whether they perceived a change in their views about the goals of bioethics as a result of their MBE program participation; 3 students felt that their views had not changed at all, while 4 felt that they had gained a broader or more complex view of the aims of bioethics. One, for example, now felt less confident that the field could convey a united front, and expressed concern about consequent loss of legitimacy, due to the explicit coexistence of multiple expert viewpoints.

Altogether, students observed that bioethics faced several challenges in achieving its goals. Responses fell, broadly speaking, into two categories: (1) that bioethics needed to do a better job of reaching its nonacademic audience, and (2) that its impact was limited due to its newness and consequent lack of authority. One student observed that bioethicists "primarily write for one another" rather than for a broader audience. With respect to the impact of bioethics, one commentator felt that the field needs to exercise stronger influence on lawmakers and the public, for example, by bringing about more equitable access to resources for health care. Another deemed that bioethics needs to "get more practical."

Several students noted the ways in which the newness of the field presents a challenge to bioethics achieving its aims. Interviewees observed that the literature of the field was not well defined, that no areas of inquiry are "saturated" yet, and that the field needed more exposure, legitimacy, and authority, to distinguish itself from "simple ethics." One student stated that bioethics is "a profession that's overwhelmed" by current events, "clueless people," and the "overpowering force of money in biotechnology;" she compared bioethics to the little Dutch boy trying to plug the hole in the dike with his finger.

CONCLUSION

This chapter examined what students are seeking and receiving by enrolling in MBE programs. While the low participation rate of students in my case study of the Center limited the power of my analysis, student interviews were informative, and together with analysis of AJOB student commentators' responses, provide some basis for drawing tentative conclusions about students' relationships to master's programs in bioethics. Overall, the responses from Letters students were consistent with the perspectives of AJOB student commentators. Both groups of students, many of whom were midcareer professionals, expressed interest in reaping both expertise and social benefits from MBE training, largely with particular career goals in mind. Some of the students at Letters also indicated that they expected an MBE to provide them with legitimacy, cachet, or an advantageous niche in their vocational pursuits.

As a whole, bioethics students at Letters University provided less glowing accounts of their program than those given by AJOB commentators, but still felt that obtaining the degree itself would be of value to them. The Jenkins report supports the contention that the variation among Letters students in their levels of satisfaction may be due to overextension of Center faculty. The undefined nature of the bioethics job market notwithstanding, MBE programs are perceived by students as being desirable and beneficial, and potentially providing them with bioethics expertise, entrance into the bioethics community, and the legitimacy of a credential. In addition to tuition revenues, bioethics centers and their faculty also gain institutional legitimacy from MBE programs. Although there is considerable consternation among tenure-track academic bioethicists about both the implications of MBE programs and students' apparent eagerness to adopt the title of bioethicist, MBE programs are unarguably contributing to the institutionalization, cohesion, and legitimacy of the field. It does not appear that the MBE is an isolated "boutique" phenomenon offering a unique training experience, but rather reflects the franchising of academic bioethics in colleges and universities across the United States. These franchises are not producing PhD bioethicists that will go on to propagate academic bioethics programs at other universities, but are chiefly developing a cadre of professionals in health care related fields who will continue to draw upon the bioethics community and its expertise to address bioethical issues in their workplaces. In this respect, bioethics is comparable to the field of higher education studies, which primarily trains administrative educational professionals who work in postsecondary institutions.

ENDNOTE

1. Similarly, many existing programs from departmental tracks have migrated into new bioethics centers.

PART II

BIOETHICS AND THE STATE

The state has had an abiding interest in science throughout the 20th century (Kevles, 1995), and following World War II, an increasingly formal and systematic relationship developed between the state and the institutions of science in the United States (Leslie, 1993; Smith, 1990). During the immediate postwar era, a social-contract approach characterized science policy, in which the state provided resources for scientific research and presumed that by priming the scientific-knowledge-production pump, the unfettered progress of science would generate useful knowledge and applications (Byerly & Pielke, 1995; Guston & Keniston, 1994). The social contract model came under fire from the antiauthoritarian sentiments of the 1960s, and in the subsequent science policy regime of collaborative assurance, the state has taken a more active role in assuring the integrity and productivity of science (Guston, 2000). After the Cold War, the state's interest in science shifted to focus on the goals of economic competitiveness (Slaughter & Rhoades, 1996).

As the state expanded regulation and funding of science, it also increasingly called upon expert advice to guide its decision making. In the executive branch, advisory committees proliferated, including a succession of bioethics advisory bodies that originated in the 1970s (OTA, 1993). In the judiciary, expert testimony began to play a larger role in advising the court on technical issues (Solomon & Hackett, 1996).

Given the close relationship between the state and science, bioethics needs to engage effectively with both institutions in order to influence the conduct of biomedical research and practice. Part II of this monograph examines how bioethics engages with the state, and the impact of that engagement on the legitimacy and logic of bioethics. In chapter 5, I argue that professional liability is an overlooked indicator of the legitimacy of jurisdictional claims in the system of professions, and I examine the implications of legal liability for the legitimacy and jurisdiction of bioethics. In order to do so, I first examine discourse on the liability of institutional review boards (IRBs) and the implications thereof for the bioethics jurisdiction; I then analyze a report developed by the American Society for Bioethics and Humanities on the liability of health care ethics consultants, where I examine the relationships between liability, expertise, and legitimacy.

In the federal executive branch, bioethics advisory bodies have become a staple of U.S. public policy for addressing social controversies such as fetal tissue research and human cloning, in spite of the limited direct impact these bodies have had on science and technology policymaking. Kelly (2003) argued that public bioethics advisory bodies serve an important tacit function as boundary organizations that stabilize the border between science and politics, thus preserving the autonomy of science from incursion by other societal stakeholders. These boundary organizations succeed in bounding and controlling controversy by constraining the set of issues and viewpoints that are addressed in discourse, and by dictating the decision-making strategy in ways that privilege the participation of some stakeholders over others and veil the intensity of controversy.

The National Bioethics Advisory Commission (NBAC), a presidential advisory body established during the Clinton Administration, was such a boundary organization. The creation of NBAC both directly legitimated bioethics, and provided opportunities for the further institutionalization of bioethics in the public-policy arena. In addition to mediating the tensions between politics and science, NBAC was also faced with negotiating the boundaries between science and ethics on the one hand, and between ethics and public policy on the other. Chapter 6 examines the boundary work performed by NBAC in its deliberations on embryonic stem cell research, and the implications thereof for the legitimacy and logic of bioethics.

CHAPTER 5

LIABILITY AND EXPERTISE: THE EMERGING PROFESSIONAL JURISDICTION OF BIOETHICS IN THE LEGAL ARENA

This chapter provides an account of how institutional review boards (IRBs) and health care ethics consultants may be held legally accountable for their expertise and decision making, and likewise proposes an explanation of how discourse on liability in the legal arena provides an opportunity to examine jurisdiction construction in the system of professions. Professional groups must meet different kinds of accountability requirements in the workplace, legal, and public arenas. In the academic workplace, bioethicists, like faculty in other fields (whether disciplinary or interdisciplinary), are held accountable for their job performance through the requirements for promotion and tenure, which may variously include measures for producing publications, grants, good public relations, and enhancement of institutional function. As discussed in chapter 2, bioethics C&Is are ultimately held accountable for establishing both resource stability and academic credibility, in order to maintain their very existence. For federal bioethics advisory bodies in the public-opinion arena, the subject of chapter 6, accountability stems from the advisory body's status as a boundary organization, which confers upon the advisory body distinct lines of responsibility and accountability to the political and scientific principals it serves. As clients and the state come to rely increasingly on a particular expertise, and consequently confer upon it greater authority or legitimacy, the providers of that expertise can become more liable for negligence and malpractice, whereby the expert is held responsible for causing injury as a result of failing to provide the expected professional standard of skilled care.

What, then, does the issue of legal liability reveal about the jurisdiction of bioethics? Concerns about the liability of IRBs and ethics consultants are closely tied to issues of accreditation, certification, and licensure, all of which are concerned with the appropriate and accountable ethical expertise of IRB members and ethics consultants.[1] Hence, discourse on liability provides an opportunity to examine the social, institutional, and legal legitimacy of bioethics expertise, since this legitimacy is constructed by relevant professional groups, institutions, and the legal system. This chapter begins by reviewing Abbott's theoretical approach to professional jurisdictional claims in the legal arena, and by making a case for further elaborating the role of the legal arena in shaping jurisdictional claims. It then analyzes the discourse concerning the expertise for which IRBs and ethics consultants may be held liable, and discusses the implications for the jurisdictional claims of bioethics.

CLAIMS OF PROFESSIONAL JURISDICTION IN THE LEGAL ARENA

One hallmark of powerful professions is the achievement, through successful lobbying for licensing regulation, of a legal monopoly on the control of work. Physicians, attorneys, and members of several other professions have secured this legal claim of jurisdiction. These claims can include the monopoly of certain activities and types of third-party payments, and the control of certain work settings and of certain kinds of language. Contests for legal jurisdiction are waged in all three branches of government, although legislative bodies have dominated the legal structuring of professions in the United States. Abbott explained that the legal definition of professional jurisdictions is a locus of what Gieryn (1999) described as boundary work:

> Boundary areas [at the margins of professional jurisdiction] are firmly delineated with formal definitions that are in fact uninterpretable in actual situations. Thus, a crucial boundary between law and psychiatry concerns the point where legal definitions of responsibility give way to psychiatric ones, the point at which the insanity defense becomes tenable. (Abbott 1988, pp. 63–64)

Legal rules were established to mark this point between law and psychiatry, but as Abbott attests, these rules have been continually disparaged due to their inapplicability to real-life circumstances.

As a result of such arbitrary formality, the formal legal jurisdictions of professions do not reflect the complex nature of actual professional existence. For example, the definitions of the beginning and end of life, perennial issues in bioethics, sit uncomfortably at the border of law and medicine.[2] Abbott (1988) contended that despite the prominence of legal matters in the mass media, the legal jurisdictions of professions reflect, rather than determine, the impact of external forces on professional jurisdiction. However, one exception noted by Abbott is that dramatic changes in jurisdiction can be imposed by the state if control of professions is exercised primarily by either of the executive or administrative branches, as it is in France, rather than by the legislative branch. In England and the United States, Abbott argued, legislatures have dominated determinations of the legal jurisdiction of professions.

I contend that in the United States, the federal executive branch dominates legal control over the professional jurisdiction of biomedical research, but that control is exercised differently than in the French model which Abbott describes. While in France, the administrative branch of government in general exercises considerable direct control over jurisdiction, professional organization, pricing, and service delivery, I argue that with respect to biomedicine the U.S. federal executive branch exercises indirect regulation of the professional jurisdiction of biomedical research through the provision of research funding aimed at state-determined research priorities, and the accompanying regulatory requirements imposed on government-funded research. It must be noted that biomedical research professionals populate state agencies that oversee research funding and regulation, revealing a tightly coupled relationship between the state and research professionals. Nonetheless, oversight of the biomedical research jurisdiction by the U.S. executive branch has tremendous and often unwelcome impact on professional work in universities, in spite of a strong research professional presence in the state. Federal regulation of publicly funded biomedical research on human subjects and live animals also influences the private sector, by generating considerable normative institutional pressure for privately funded researchers to voluntarily comply with federal standards.[3] IRBs, which shall be discussed shortly, are a cornerstone of governmental regulation of biomedical research.

Beyond state licensure and funding, the legal arena provides other opportunities for the legitimation of professional expertise and the staking of knowledge claims, although these opportunities do not confer as much jurisdictional power. Firstly, the judicial forum legitimates certain types of expertise by utilizing professionals as expert witnesses in court cases. Analysis of the expert testimony and its influence on court decisions can reveal strengths and weaknesses of jurisdictional claims.[4] Secondly, the legal basis for negligence claims involves elements of expertise, and examining the nature of important expertise elements provides significant clues regarding the strengths and weaknesses of a professional jurisdiction. Tort claims and court decisions indicate what elements of expertise are considered important by the audiences of the profession in question, and contribute to an explanation of the influence of those audiences on the jurisdiction of that profession, by revealing what expertise the profession is liable for.

While these elements of professional jurisdiction claims in the legal arena are weaker than the direct legal controls of jurisdiction secured by powerful professions, they serve as indicators of professional jurisdictions in the making, such as that of bioethics, as they emerge in a litigious society. With increased power comes increased responsibility. The legal formalization of professional jurisdiction confers, and defines, not only some degree of jurisdictional control, but also liability for malpractice in that jurisdiction. It is therefore prudent for professional groups to stake out their jurisdictions carefully, as far as is possible.

How does liability serve as an indicator of the bioethics jurisdiction? How does accountability in the legal arena relate to expert accountability in the other arenas, and to other sources of legitimacy? To answer these questions, it is useful to consider the liability of universities, their IRBs, and biomedical faculty engaged in research, because bioethicists may serve these other liable parties in an advisory capacity. When court is adjourned, who is held responsible for negligence (i.e., is it the investigators, the IRBs, or the bioethicists), why, and with what implications for the jurisdiction of bioethics?

THE LIABILITY OF IRBS

Institutional Review Boards, or IRBs, are committees designed to review research involving human participants, to ensure ethical conduct and protection of participants. IRB review is required for any research funded by the U.S. Department of Health and Human Services or certain other federal agencies, and has also been widely adopted for privately sponsored research. The U.S. Food and Drug Administration (FDA) requires IRB review of any research involving the administration of investigational drugs or medical devices to human participants. IRBs have the authority to approve, require modifications of, or disapprove research projects. There are specific requirements for the constitution of IRBs, calling for the participation of both scientists and nonscientists, demographic and cultural diversity, and special representation for the review of studies involving vulnerable populations such as prisoners.

IRB member participation in protocol review is governed by conflict of interest rules.

Although well established, the IRB mechanism has a dubious reputation similar to that of the FDA: Both institutions have been widely and simultaneously criticized both by those who find these institutions too lenient, and by others who find them too restrictive. Decades of tweaking by policy makers have failed to diminish either complaints about IRBs or suggestions for their improvement.[5] IRBs have been around in one form or another since 1966, when the U.S. Public Health Service required that any research it sponsored be reviewed by a committee of associates, as part of providing assurance of compliance with human research regulations. Federal research regulations underwent several revisions over the next 25 years, and in 1991, major revisions resulted in what is known as the Common Rule (45 CFR 46), which was adopted by 16 federal agencies and departments.

IRBs received renewed attention beginning with the 1998 release of the report *Institutional Review Boards: A Time for Reform* by the Office of the Inspector General (OIG) in the Department of Health and Human Services (DHHS). The OIG report sparked congressional hearings, the revision of regulations by DHHS, and increased oversight by the Office of Human Research Protections, housed in DHHS. Scrutiny of IRBs was further heightened by the death of teenager Jesse Gelsinger in a Phase I gene therapy trial at the University of Pennsylvania in September 1999. The tragedy raised concerns about both conflict of interest and appropriate informed consent, led the FDA to temporarily suspend all of the university's gene therapy trials, and resulted in a lawsuit by the Gelsinger family. The year 2001 witnessed two more human research incidents, this time at Johns Hopkins University (JHU). First, JHU was sued by the family of Ellen Roche, a healthy subject who died in a Phase I clinical trial examining the use of the blood pressure medication hexamethonium as a treatment for asthma. The federal Office for Human Research Protections (OHRP) temporarily suspended all federally funded human subjects studies at Hopkins facilities. Subsequently, JHU issued an angry response[6] and asserted, "OHRP's action seems to us to be an extreme example of regulatory excess" (JHU, 2001). The Kennedy Krieger Institute, an affiliate of JHU, was also sued in 2001 for negligent sponsorship of lead abatement studies and inadequate informed consent from parents of children exposed to harmful levels of lead.

In response to these and other related events, an IRB accreditation program was created in 2001 by the Association for the Accreditation of Human Research Protection Programs (AAHRPP), which was newly founded by seven national professional and postsecondary organizations: the Association of American Medical Colleges, the Association of American Universities, the Consortium of Social Science Associations, the Federation of American Societies for Experimental Biology, the National Association of State Universities and Land-Grant Colleges, the National Health Council, and Public Responsibility in Medicine and Research (PRIM&R). Prior to that, the Applied Research Ethics National Association, a division of PRIM&R, established the Council for Certification for IRB Professionals, which made competency-based testing for certification, available in 17 states, the District of Columbia, and four Canadian cities as of October 2005. The availability of testing for certification expanded to include 49 states (all but Rhode Island) and a total of 8 Canadian cities by April 2008. IRB accreditation and professional certification are voluntary, and do not, at this time, confer the same prestigious ethical distinction associated with the accreditation of animal research facilities by the Association for Assessment and Accreditation of Laboratory Animal Care (Resnik, 2004). However, as of October 2005, 27 institutions in the United States, including 4 private IRBs, 1 federal institution (Hunter Holmes McGuire Veterans Affairs Medical Center in Richmond, VA), and several hospitals and prominent research universities (including Baylor, Johns Hopkins, the University of Iowa, the University of Minnesota, and Dana Farber/Harvard Cancer Center), had received full or qualified AAHRPP accreditation.[7] In March 2008, AAHRPP issued a press release reporting that a total of 107 organizations had received accreditation, as 15 new organizations became newly accredited, including the first state health department, Florida (AAHRPP, 2008). Outside the United States, 1 institution each in Canada, South Korea, and Singapore have successfully applied for accreditation from AAHRPP. The hope that accreditation can help prevent research litigation will undoubtedly serve as an incentive for participation by even more research institutions.

It is worth noting that investigators, and not just study participants, may take legal action against IRBs, although it appears this will be less common. In *Halikas v. University of Minnesota*, a researcher sought a preliminary injunction against the university's IRB for publicly announcing why it had halted his research. The court denied the injunction, affirming that it would impair the public-protection function of the IRB, and that this harm was greater than the questionable harm claimed by the plaintiff. The court also found that the plaintiff had received due process in the IRB's investigation and actions, confirming that the IRB had abided by federal human research regulations and the university's general assurance agreement, both of which provide for due process, but that the IRB was not bound by the university tenure code or by hospital bylaws (Maloney, 2003).

IRBs could be a target for future litigation by investigators, particularly where failures of due process can be documented. Research industry stakeholders have proffered both defensive and offensive reactions to IRB oversight, exemplified respectively by Dimond's and Alvarez's (2004) preventive advice article, "Investigator-Initiated Research: Proceed with Caution," and the defiant "Researcher's Bill of Rights," presented by Perlstadt (2004) at the Michigan State University Center for Ethics and Humanities, and published in the *Medical Humanities Report* (Life Sciences, Winter 2004 issue). With research negligence lawsuits on the rise, however, one would expect the research enterprise, including universities, hospitals, biomedical technology firms, and individual investigators, to be more concerned with their own liability than with filing claims against IRBs.

The lead prosecuting attorney for the Gelsingers, Alan Milstein, has filed 20 lawsuits against universities, hospitals, investigators, and IRB members (Lockwood Tooher, 2005). Five of these cases, filed since 2001, named IRB defendants. Individual IRB members were named as defendants for the first time in 2002, in *Robertson v. McGee*, again filed by Milstein (*Robertson v. McGee*, 2002). The case was filed against the University of Oklahoma IRB on behalf of 12 melanoma patients who participated in a Phase I cancer vaccine trial, alleging therapeutic misconception against the investigator, and failure to provide adequate continuing review. This development was cause for alarm among IRB members nationwide, most of whom serve on a voluntary basis for their home institution or other institutions.[8]

Many clinical trial lawsuits have been settled out of court, and so, it remains to be seen what guidance will emerge after the courts have ruled on more of these cases. There have not yet been any successful negligence cases filed against IRBs or their members. However, Resnik (2004) argued that the number of lawsuits filed against IRBs is likely to increase, due to a general increase in research litigation, and the incentives that large settlements provide to plaintiffs' attorneys. Likewise, Icenogle (2003) predicted an increasing risk of liability for IRBs, citing as factors in this increase the high pressure to approve studies and bring in research dollars, and the growth of genetic technology and its attendant gamut of concerns and unknowns.

Mello, Studdert, and Brennan (2003, p. 43) agreed that "prospects for the growth of tort litigation in human subjects research are extremely favorable," due to developments including the diversification in number and type of claims filed by plaintiffs, the expanding array of defendants named in cases, and the growing use of class action suits. With respect to the diversity of claims filed, the introduction of fraud claims is particularly noteworthy, because of their powerful effect on jurors and potential for large damage awards, including punitive damages. The defendants named in suits now include not only investigators, hospitals, universities, and pharmaceutical sponsors, but also top university officials, individual IRB members, and ethics consultants, causing the cases to attract more media attention. Clinical trials litigation is particularly amenable to class actions, because of the large number of clinical trial participants and the commonalities shared by them. Class actions pose advantages for plaintiffs' attorneys, who can achieve economies of scale by joining the forces of plaintiffs' firms, and earn higher contingency fees through a potentially large number of plaintiff awards.

Other factors arguably make research litigation cases more attractive to plaintiffs' attorneys than traditional medical malpractice cases (Mello et al., 2003). Firstly, it is easier to question the intent of investigators than of physicians, who are generally perceived as intending to help the patient. Secondly, the legal standard of care relevant to negligence and malpractice cases has a different basis for research than for medical practice; in the latter, professional standards of care are typically invoked and provide straightforward guidance for the court to determine the required standard of care whereas there is no universal professional standard of care in research for the court to rely upon. The nature of the legal standard of care is key to the relationship between expertise and liability, as I will discuss shortly.

As a small number of research lawsuits have started working their way through the courts, several scholars have begun to reflect on the nature of research malpractice as it becomes more distinct from medical malpractice. Resnik (2004) explained the legal duties of IRB members and their likely liability for negligence based on existing case law, and discussed the viability of some legal defenses for IRBs. Morreim (2004) observed that research-related tort claims have been unreflectively subsumed by the courts under medical malpractice law, and so, the body of case law does not yet adequately reflect the distinctive aspects of research-related injuries. She argued that tort doctrine needs some adjustment in order to treat both research participants and the various parties in the research enterprise fairly in clinical trial lawsuits. Shaul and colleagues (2005) noted that it may not be obvious to health care professionals that the legal standards for informed consent are higher for research than for medical treatment; the informed-consent process for research must reveal all known risks, including remote risks that could have severe consequences.[9] The authors also raise the possibility that IRBs could be faced with lawsuits filed by the federal government. This risk is inferred from the Gelsinger case, where the defendants settled not only with the Gelsinger family, but also with the federal government, which filed a separate civil action.

In the present climate, both investigators and IRBs skimp on human subjects protections and review of research protocols at their own peril. Although the IRB responsible was not named as a defendant, the *Grimes v. Kennedy-Krieger Institute* decision on the defendant's lead abatement study set a precedent in which the court effectively overruled the IRB's approval of the study as ethical research. The court concurred with the plaintiff that the research was unjustifiable because the harm of exposing children to hazardous levels of lead outweighed any possible benefit, contrary to the prior conclusions reached by the Johns Hopkins IRB. "In essence," Mello et al. (2003) concluded, "the court replaced the expert judgment of the IRB with its own judgment of the risk-benefit ratio, suggesting that *neither the parents' consent nor the IRB's approval of the protocol would make the researchers immune from liability* [italics added]" (p. 42). The ability of IRBs to protect research institutions and investigators from litigation has clearly been circumscribed, and the scope of judicial review and regulation of human subjects research has increased to include not only investigators' conduct, but also the deliberations of IRBs.

Plainly, IRBs are not immune to scrutiny by the courts, anymore than they have escaped criticism from the parties meeting each other across the informed-consent form.[10] On what basis, then, are IRBs legally liable for research-related injuries? The next section encapsulates Resnik's account of legal liability for negligence, and indicates how determining the most difficult element of a negligence claim against IRBs—namely, establishing standard of care—presents an opportunity for bioethics to legitimate its jurisdiction in the legal arena.

IRBs and the Legal Concept of Negligence

Resnik (2004) described six necessary elements to a legal claim of negligence,[11] and their implications for IRB liability: duty, standard of care, breach of standard of care, cause-in-fact, legal cause, and damages (harm or injury). At least three legal theories could be used to establish the *duty* of IRBs to research participants. This legal duty may be based on the legal requirements of federal research regulations, on the theory of reasonable supervision of researchers, or on contractual duties to participants as third-party beneficiaries. Resnik (2004) cited two cases (*Kus v. Sherman Hospital* and *Gregg v. Kane*) that established the legal duties of IRBs under federal regulations. Although IRBs were not named as defendants in either of these cases, if they had been, Resnik finds that the plaintiffs would have had valid negligence claims.

There are various legal bases for establishing the *standard of care* implied by a legal duty. Several of these bases could be applied to the duty of IRBs, including the reasonable person standard, professional standards, and international standards, but one that seems particularly appropriate in the case of IRBs is a statutory standard, whereby the law itself sets the standard of care. Several federal regulations, most notably the Common Rule, as well as some state laws, govern the structure and function of IRBs, and the conduct of IRB members. Two cases (*Grimes v. Kennedy Krieger Institute* and *Whitlock v. Duke University*) provide precedent for recognizing the federal regulations as the appropriate standard of care in research, and for applying that standard to IRBs. There are, however, some challenges associated with establishing a clear legal standard of care, an issue to which I will return after outlining the remainder of the legal elements of negligence for IRBs.

Cause-in-fact can be legally established as either a "but for" cause or a substantial factor. In the first instance, the court must find that harm would not have occurred *but for* the defendant's actions, which fall short of the standard of care. In the instance of a substantial factor, more than one defendant perpetrates a single negligent act, whether independently or jointly, to which each of the defendants contributes a substantial factor. Resnik (2004) argued that cause-in-fact for IRB negligence could readily be established for failure to perform either the initial or continuing research review functions of IRBs. Individual members of the IRB may be implicated if the court finds that individual members of the IRB, such as the chair and the primary reviewer, assume different responsibilities for a harm; such individual responsibility could be tied to the professional roles of members, such as those of physician members and ethicist members, or to functional roles, namely those of IRB chair and the primary reviewer.

A viable negligence claim must also demonstrate *legal cause*, or proximate cause. Legal responsibility for injury often rests on the general proximate cause rule that the harm is a reasonably foreseeable, but not necessarily probable, consequence of the defendant's conduct. Resnik (2004) envisaged that in most cases, establishing legal cause of negligence by an IRB would be straightforward.

The *damage* resulting from negligence can include physical, psychological, and economic harm. The court may award damages to the plaintiff for these harms, as well as punitive damages for malicious or wanton behavior by the defendant. All of these damages can occur as a result of negligence in research on human subjects. A high-profile example of this is the settlement between the U.S. government and the survivors and heirs of the Tuskegee Study, as noted in chapter 1, which included damages for physical and psychological harm.

It is clear from Resnik's (2004) analysis that IRBs, and their individual members, could readily be found liable for negligence in the oversight of research with human subjects, especially when clear violations of federal regulations have occurred.[12] However, he contends that establishing the legal standard of care for IRBs may prove to be a difficult task in many cases, for reasons which shall be explained in the next section. This difficult task is one that falls under the purview of bioethics, and may provide an opportunity for bioethics to solidify its professional jurisdiction, as more suits charge IRBs with negligence and make it to trial.

Determining the Standard of Care for IRBs: A Role for Bioethics Expertise

Of all the six necessary elements to a legal claim of negligence, Resnik (2004) explained that the most difficult aspect of demonstrating IRB negligence is establishing a breach in the standard of care. While federal regulations address the standard of care for research review and monitoring, there is a great deal of disagreement about the meaning of key terms in the regulations, and about how the regulations should be interpreted and applied. Attorneys, ethicists, researchers, government officials, and patient advocates dispute the meaning of regulatory terms such as "risk," "minimal risk," "research," and "therapy," and argue about how to apply the regulations for the purpose of determining appropriate risk-benefit ratios; valid informed consent; the ethical use of placebo controls; and appropriate review, reporting, and monitoring of research protocols. Accordingly, a defendant might appeal to ambiguity in the regulations, arguing that they therefore do not apply to the IRB conduct in question.

One approach to establishing the legal standard of care for IRBs is the application of international standards. Prior to the Gelsinger case, most clinical trial lawsuits alleged medical malpractice, but Milstein, who represented the Gelsingers, has been credited with developing a new legal approach that cites the Nuremberg Code, an ethics guidance document adopted internationally in response to research atrocities committed in Nazi concentration camps, as well as the Belmont Report (Lockwood Tooher, 2005; National Commission, 1979). In developing this approach, Milstein has attempted to make various arguments establishing a right to be treated with dignity, and furthermore that violations of that right fall within federal jurisdiction.[13] Resnik (2004) pointed out that three such cases citing the Nuremberg Code have been rejected by the courts on the grounds of a lack of federal jurisdiction, but conceded that international standards could still be useful in establishing the standard of care for research participants in state court cases. It is possible, however, that U.S. institutions engaged in international research, and their IRBs, may soon be found liable by U.S. courts for violations of international standards of care under the Alien Torts Claim Act (Shaul et al., 2005).

Mello and colleagues (2003) contended that one of the reasons research litigation may be more attractive to injury attorneys than medical malpractice suits is because of the different legal bases for discerning the standard of care. Whereas the professional standard of care in medical practice can be relatively straightforwardly evidenced by appropriate expert testimony, the standard of care in research is drawn from federal regulations, international guidelines, and the standard of what a "reasonable" IRB would require. The authors note,

> *The use of a reasonableness standard gives judges and juries wider leeway* [italics added] than a custom-based [i.e., customary professional medical practice] standard *in determining what should be required of IRBs* [italics added] and investigators. In practice, this standard is tougher on defendants, who cannot invoke an 'everybody does it' defense. (p. 43)

As discussed previously, the need for IRBs to exercise considerable judgment and discretion in determining the standard of care makes the court's task of discerning a "reasonable" standard of care a complex matter. Mello and colleagues (2003) contended that public concern about research safety will likely result in determinations of IRB reasonableness that favor plaintiffs.

It is the responsibility of the court to interpret the legal standard of care entailed by human subjects regulations. The court should first attempt to discern legislative intent, Resnik (2004) argued, by examining the history of the statute and its language. Such investigation should include testimony relating to the history of human research abuses and key documents that have shaped the regulations, such as the Belmont Report (National Commission, 1979). If the courts are still unable to resolve any ambiguity, they may defer to reasonable and persuasive interpretation and to the expertise of agencies authorized by Congress to implement the regulations, including the National Institutes of Health, the Office of Human Research Protections, and FDA. Resnik (2004) envisaged that even after these steps, in some cases a clear interpretation will still be lacking, because

> Even if one can determine the meaning of key terms and phrases, questions will arise concerning their application to particular research studies. The regulations require IRBs to make many different decisions that involve a great deal of *judgment and discretion...even experienced IRB members or federal officials may disagree about risk/benefit assessments and the reasonableness of risks in relation to benefits* [italics added]. (pp. 153–154)

I argue that when IRB members or federal officials disagree, the problem of interpreting the regulations and assessing whether appropriate judgment and discretion has been exercised by IRBs reflects a jurisdictional opportunity for bioethicists, whose expert testimony may be sought by plaintiffs, defendants, or the court itself, in order to clarify interpretation of the human subjects regulations. Furthermore, bioethicists can and do provide education and consultation on research ethics for investigators and IRBs seeking to prevent unethical research, scandal, and litigation. Indeed, one jurist affirms in the *Washington Law Review*, "expert testimony should establish the standard of care for research malpractice, and IRB approval should be a partial defense" (Jansson, 2003, p. 229).

In cases of professional negligence, experts may testify based on their knowledge of a particular profession, such as medicine. Resnik (2004) contended that the courts would likely admit expert testimony, from bioethicists, for example, in order to establish a *professional* standard of care for IRBs. However, whether the courts would *require* such testimony depends on whether the ordinary person can discern the standard of care in a particular case; only if they cannot does the conduct in question become a matter of professional negligence, requiring expert testimony. Since ordinary people (i.e., unaffiliated nonscientists) are required by the federal regulations to be represented among IRB members, Resnik (2004) expected that a court could readily find an ordinary person able to understand the expertise exercised in the regular duties of an IRB member.

The problem of establishing the legal standard of care for IRBs, which may be approached by interpreting statutes, or by appealing to international or professional standards, is clearly within the jurisdiction of bioethical expertise, which provides multidisciplinary tools for illuminating various social and semiotic nuances that arise in the course of interpreting and applying the various standards. Bioethicists not only can provide expert testimony to assist with discerning the legal standard of care for research participants, they also can provide education and consultation to investigators and IRBs in order to better prevent negligent research and consequent lawsuits.

Furthermore, if IRB accreditation becomes standard practice, and IRB members, including community and other nonscientist members, are regularly certified, alleged research negligence could eventually come to be recognized as a matter requiring professional testimony. Given the elite organizations and institutions that have supported and sought IRB accreditation, I anticipate that accreditation of IRBs will soon be expected of all research institutions. These observations suggest a considerable opportunity for bioethics to acquire greater prominence and legitimacy in the legal arena, as the number of research-related tort claims increases. However, bioethics faces two challenges to reinforcing this jurisdictional claim (both of which were discussed previously in chapter 3), namely, a reluctance to claiming moral authority, and the tension in the relationship between academic bioethicists and IRBs.

Seeking control of work by lobbying for licensing regulation is only one of several ways in which claims of professional jurisdiction may be staked in the legal arena. The previously presented summary of the liability of IRBs reveals that the federal regulations governing IRBs have created a new professional task, that of establishing the appropriate standard of care for IRBs. Interpreting and applying the Common Rule requires complex discretion and judgment on the part of IRB members, and frequently results in disagreement among experts about the meaning and application of key terms. To the extent that bioethicists or other professionals are able to remedy these problems, they may stake a jurisdictional claim. In the current environment of research litigation, IRBs, as well as courts, will seek out and legitimize expertise that usefully illuminates the standard of care to which IRBs are held, for the purposes of respectively preventing or establishing judgments of negligence and malpractice (or alternately, avoiding expensive settlements). Universities and IRBs are highly motivated to avoid the negative publicity, research suspensions, and economic repercussions of research lawsuits.

Another implication of research torts for the professional jurisdiction of bioethics, as suggested previously in this chapter, is the incentive created to pursue accreditation as a protection against lawsuits. Accreditation adds a new layer of social organization to research ethics, promotes institutional isomorphism, and clarifies the expertise and performance standards for IRB function and membership, both defensively for the research enterprise, and offensively for plaintiffs in research lawsuits. All of these effects of accreditation will arguably strengthen the legitimacy of bioethics by organizing and standardizing ethical expertise and affirming its utility for clients.

While bioethicists often face difficulties serving as IRB members, they can impact research ethics by contributing to the ethical discourse which many IRB members and investigators draw upon to design and review research protocols. The dialogue between IRBs and bioethicists is not entirely hostile. Recent annual PRIM&R conference programs have featured several prominent bioethicists as presenters in plenary and concurrent conference sessions, who spoke to an audience primarily comprising research administrators, IRB administrators, IRB members, and government research oversight officials. It is also worth noting that the universities which have thus far been sued for negligent research are all elite research institutions, where one might also expect to find the most cantankerous relationships between bioethicists and IRBs, given the tremendous pressure to procure grants and publish first-tier journal articles. Fear of litigation may well drive IRB members at these institutions into the arms of bioethics.

The legal liability of a professional jurisdiction is a clear indicator of socially valued expertise, and establishes those elements of knowledge and skill that professionals need to master and use carefully. These elements can include not only specialized technical skills and knowledge, but also procedural knowledge and a commitment to professional virtue, which, in the case of biomedical science, is closely linked to sensitivity toward the expectations and interests of clients, and in particular, research participants.

THE LIABILITY OF ETHICS CONSULTANTS

The 1998 American Society for Bioethics and Humanities (ASBH) report, *Core Competencies for Health Care Ethics Consultation* (discussed in chapter 3), defined health care ethics consultation as "a service provided by an individual or a group to help patients, families, surrogates, health care providers, or other involved parties address uncertainty or conflict regarding value-laden issues that emerge in health care," and notes that ethics consultation includes the domains of both clinical ethics and organizational ethics (ASBH, 1998, p. 3). Consistently with previous observations regarding the reluctance of bioethicists to adopt a moral stance, the ASBH (1998) report rejects both an *authoritarian ethics consultation* approach, in which consultants are the primary decision makers, and a *pure facilitation* approach, where the consultant only seeks to coordinate a consensus among involved parties, without clarifying the social, legal, and institutional dimensions of the moral choice. Instead, the report advocates an *ethics facilitation* approach, which incorporates both "identifying and analyzing the nature of the value uncertainty and facilitating the building of consensus" (p. 6). As shall be discussed, the approach chosen by ethics consultants has implications for their liability.

Ethics consultation is nearly as old as bioethics itself, dating back to the late 1970s. The Society for Bioethics Consultation was established in 1985, and eventually merged with the Society for Health and Human Values and the American Association of Bioethics in 1998 to form ASBH. In 1992, the Joint Commission on the Accreditation of Healthcare Organizations (JCAHO) presented new ethics criteria in its *Accreditation Manual for Hospitals* (JCAHO, 1992), which likely generated increased demand for ethics consultants (McCarrick, 1993). In a special issue of JCAHO's periodical *Quality Review Bulletin*, published in 1992, the same year new ethics criteria appeared in the JCAHO's accreditation manual, the potential for future litigation against ethics consultants is discussed (Brennan, 1992).

Serious concern about the liability of ethics consultants can be traced back to 1986, when a hospital ethics committee was named, as a whole and as individual members, as a defendant in *Bouvia v. Superior Court*, after supposedly directing physicians to insert a nasogastric tube in the plaintiff against her wishes. The suit was dropped after the plaintiff secured an injunction to have the tube removed, but ethics consultants were made permanently aware of their potential liability (Poland, 1997; ASBH, 2004). Over 10 years passed before a high-profile lawsuit named prominent bioethicist Arthur Caplan among the defendants sued by the family of Jesse Gelsinger, following his death in a gene therapy trial. Caplan had advised the investigators to conduct the study with healthy adults rather than sick infants; ultimately he was dropped from the suit. Since the 1980s, consultation with ethics committees has been outpaced by consultation with individual ethics consultants, which has undoubtedly increased the self-perceived vulnerability among consultants (Sontag, 2002).

In October 2003, Jonathan Moreno, then president of ASBH, commissioned an ASBH Task Force to produce a report on ethics consultation liability for members of the ASBH. In 2004, ASBH released a task force report on ethics consultation liability, to provide its membership with an overview of the issues and information on insurance options (ASBH, 2004). The task force and report were spurred by an impetus from the Board of Directors to expand ASBH activities to better serve the needs of its professional members, rather than simply to organize annual conferences; hence, the liability of ethics consultants was identified as an area of increasing concern (Moreno, 2004).

Moreno asked Gerard Magill to chair the task force, and Magill selected other members in consultation with Moreno and many scholars in the field, emphasizing the need for representation from a wide range of expertise and constituencies (see appendix C of ASBH, 2004). Ultimately 10 other members, including 5 academic center/institute directors, were listed as authors of the report. Due to time constraints, the task force members served as an oversight committee for the report, which was primarily researched and written by the task force chair. Members provided, at their own discretion, direction, critique, and evaluation, and signed the report in support, in dissent, or with qualifications.

While the report may thus be construed as considerably the work of one man, it was also arguably shaped and endorsed by other leading bioethicists on the task force, and its findings drew upon extensive review of the existing literature (over 180 works were cited), court cases, and direct inquiries made to "a large number" (ASBH, 2004, p. 33) of insurance companies. To the best of my knowledge the Task Force report

(ASBH) is the only comprehensive account of its kind on liability and ethics consultation.

The following section of this chapter uses the report to examine the relationship between expertise and liability in ethics consultation, and the implication thereof for the jurisdictional claim of bioethics, particularly in the legal arena. It shows that in the system of professions, the ways in which other professional groups and the state perceive the professional status of ethics consultants (and by extension, bioethicists in general) shape expectations about the liability of those ethics consultants, and consequently about their jurisdictional claim. The account begins by examining, at length, the perspective of the task force on the expertise and liability of ethics consultants, and subsequently reflects on some key attitudes of health care professions, insurers, and the state toward ethics consultation that emerge from the report. The account concludes by considering questions about the construction of jurisdictional claims and the sources of legitimacy for those claims.

Ethics Consultants: An Uncertain Professional Group

As chapter 3 discussed, bioethicists are caught in a tension between an ideology of *social trustee professionalism*, grounded in the ideal of service to the public good, and an ideology of *expert professionalism*, legitimated simply by specialized authority over a defined area of formal knowledge (Brint, 1994). This ideological tension poses a particular challenge to the legitimation of bioethics, which, in spite of its public-service-oriented roots, faces pressures to seek legitimacy through expert professionalism, pressures now exacerbated by concerns about liability, as shall become evident shortly. The *Ethics Consultation and Liability* report argues that the professionalism, or the status of ethics consultation as a profession, influences both legal liability and insurability, but the report also observes considerable uncertainty among ethics consultants as to the professional status of ethics consultation (ASBH, 2004). Much of the uncertainty regarding the professional status of bioethics stems from the diversity represented within the practice of ethics consultation. One of the sources of this diversity is the array of professional backgrounds that ethics consultants represent. Hence, the report advises consultants to "speak frankly with patients and clients about their professional backgrounds, their limitations, and preferred methods of analysis" (p. 10), which can help to reduce liability. The practice contexts of ethics consultants are another source of diversity; the report considers ethics consultation in the contexts of clinical ethics, research ethics, and corporate health care ethics. The report also describes the variety of social structures and roles that characterize the work of ethics consultants. For example, some consultants work as part of a health care team, while others work independently; consultants may fill various roles including advisor, educator, counselor, advocate, negotiator, case manager, or some combination of these and other roles.

As mentioned previously, a prior ASBH task force developed the report *Core Competencies for Health Care Ethics Consultation* (1998), which, in spite of the wide scope of consultants' practice, established a set of recommended skills and core knowledge for ethics consultation. Chapter 3 discussed the fact that the task force promoted the *voluntary* adoption of these competencies, rejecting competency-based accreditation or certification. The core competency skills fall within the categories of ethical assessment skills, process skills, and interpersonal skills. Core competency knowledge comprises knowledge of ethical reasoning and theory, common ethical issues and concepts, health care systems, clinical context, local health care institutions and their policies, relevant ethical and professional codes of conduct, accreditation guidelines, and health law.

The heterogeneity of consultants' training, roles, and practice, as well as the voluntary nature of the core competencies, make it a challenge to characterize the expertise of ethics consultants, let alone assess their liability for that expertise. The *ASBH Task Force Report on Ethics Consultation and Liability* (2004), acknowledged debates about the objectivity of ethics expertise and its relationship to virtue, concluded that essential ethics expertise appears to encompass "appropriate knowledge of general principles and theories of ethics; analytical skills such as discernment and insight; and strength of will that prevents the ethicist from taking easy ways out of complex dilemmas" (p. 16). Consultants' authority to provide this expertise, the report suggests, can derive from *social authorization*, in which consultants address difficult ethical problems found intractable by others; from *institutional authorization*, where the organizations confer authority by utilizing consultants; and from *legal authorization*, through, for example, laws addressing ethics consultants' participation in decision making in patient care, or through court decisions that address ethics consultation or use ethics consultants as expert witnesses.

After establishing some fundamental characteristics of expertise in ethics consultation, the report turns to liability criteria relating to that expertise. The task force argues that although ethics consultation lacks several hallmarks of typical professions, such as standard training requirements, power of self-regulation, state licensure, and a code of ethics, it seems to be moving "toward greater professional recognition" (ASBH, 2004, p. 19), which will impact insurance companies' assessment of liability risks. The three categories of liability assessment criteria identified include competence, role types and activities, and authority, drawn from the report's previously cited analysis of the nature of ethics consultation expertise.

The report envisaged, not surprisingly, that insurance underwriters will find consultants who have little or no formal training to be at higher risk for liability. Ethics consultants who serve as generalists rather than specialists for particular settings (e.g., intensive care) are also argued to be at greater liability risk, because there is far more literature for a generalist to keep current with. The report also contends that consultants who serve in a facilitation or education role will have a lower liability risk than those in an advice or intervention role, where the consultant takes more direct responsibility for decision making. Professional authority, as the final criterion, is found to be proportional to the risk of liability.

In spite of the broad scope of clinical ethics consultation practice, the ASBH task force identified characteristic expertise elements for which they felt ethics consultants could likely be held accountable in lawsuits. The next section of this chapter turns to the perspectives of other professional interests in the system of professions with regard to the expertise and liability of ethics consultation.

Other Professions: Complex Relationships in a Complex Setting

The relationship of health care ethics consultants to other professional groups is complex, partly due to the fact that most ethics consultants have formal training in one or more of the health care related professions (including medicine, nursing, social work, clinical pastoral care, and health law). Accordingly, ethics consultants may have some liability coverage as a result of membership in other professional associations related to their training, such as the American Medical Association, for example. However, the report acknowledges that, "Because of the complexity in tracking all related professions," the liability task force elected not to examine "whether these [other professional] associations might regard ethics consultation as part of their professional expertise" (ASBH, 2004, p. 3).

The relationship between ethics consultants and health care professionals is also made complex by the nature of the clinical setting. Several psychosocial, political, and moral factors come into play, such as the attitudes and institutional power of attending medical staff, the problems of the origin of consultants' moral authority, pervasive poor communications in hospitals, the availability of consultation services, and the standing of the consultant's recommendations (including whether they are included in the patient's medical record) (Frader, 1992). Several studies found that a majority of a sample of physicians and nurses reported ethics consultation services to be useful for purposes such as clarification of ethical issues, education of the medical team, increased confidence in medical decision making, and improved patient management (LaPuma, Stocking, Silverstein, DiMartini, & Siegler, 1988; McClung, Kamer, DeLuca, & Barber, 1996; Orr & Moon, 1993). Ethics consultations have also been associated with reduced health care costs and decreased use of inappropriately prolonged treatment (Heilicser, Meltzer, & Siegler, 2000; Schneiderman et al., 2003).

However, one study found that ethics consultation services resulted in significant changes in patient management only 36% of the time (Orr & Moon, 1993). Another study found several significant barriers to the use of ethics consultation by medical residents (including power inequities within the medical hierarchy), and negative perceptions about the use of ethics consultation (including concerns about relinquishing decision-making authority, fear of delays in provision of care, doubt regarding the usefulness of ethics consultation, and the fear of confronting attending physicians) (Gacki-Smith & Gordon, 2005). Many health care providers may not understand the role of ethics consultation, and may therefore be wary of seeking it. Sometimes consultation is viewed as helpful, but other times it is viewed as a potential hindrance to medical decision making, or as an intrusion into the physician-patient relationship.

As mentioned previously, since the 1980s, consultation with individual ethics consultants, rather than consultation with ethics committees, has become increasingly more common. One reason for this is convenience. Hospital ethics committees cannot be summoned immediately and they only address one case at a time, whereas individual consultants can each be quickly called upon to address different cases (Sontag, 2002). Another likely reason for the popularity of using individual ethics consultants is that physicians are accustomed to consulting with medical specialists on a one-on-one basis, and are probably also more comfortable seeking ethics consultation from an individual than from a group (Sontag, 2002). As ethics consultants increasingly provide services not only to health care professionals but also to patients and their families, consultants may become more liable for the expert advice they dispense than when they worked more exclusively with health care professionals.

Insurers: Lukewarm about Ethics Consultation
The liability task force report observes that most academic ethicists will already have some sort of liability insurance from their institution, and consequently, it focuses on the liability of individual consultants when they provide services beyond the purview of their institutions' liability policies. However, the Task Force report observes "there appears to be

an ambivalence among insurance companies about recognizing ethics consultants as a profession" (ASBH, 2004, p. 5). The final section of the report, providing results of the task force's investigation of insurance options, stated, "Preparing this section has been quite a challenge, not least from the widespread lack of interest among insurance companies about offering a professional liability policy for ethics consultants" (p. 29). This lack of interest was explained variously by insurers in terms of a requirement for a minimum $5000 annual premium (the task force chair had suggested $500 per annum in his inquiry to the insurer who proposed this explanation); ineligibility due to perceived high risk (no reasoning was given for this explanation); and ethics consultants' lack of compatibility with any of the professional groups (e.g., medical professionals) covered by company programs.

If the task force is correct in asserting a relationship between professionalism and liability, it appears that the insurance industry finds ethics consultation lacking in professional status. It is also appears that the insurance industry is still figuring out where ethics consultation fits in the professional landscape; perhaps insurers are unfamiliar with the services ethics consultants provide, and in the absence of any case law concerning ethics consultation, they find no basis for assessing liability. An inquiry with one insurance company, Healthcare Providers Service Organization, led to the finding in the report that this insurer "<u>cannot</u> [emphasis original] offer ASBH members a professional liability policy *because ethics consultants in health care do not fit into any professional group covered* [italics added]" by their program (ASBH, 2004, p. 38). Hartford Financial Products, on the other hand, which does not provide policies for medical professionals, held a different view; the report notes that "the underwriter considered ethics consultation in health care to be within the category of medical services, even though ethics consultants typically do not have formal training in medicine" (p. 38). These two companies appear to disagree on whether ethics consultation constitutes a medical service, with the first putatively observing an important distinction in professional function, and the second affirming common institution or workplace setting as a key to eligibility. If and when a critical mass of lawsuits have been filed against ethics consultants, creating a sufficient demand for liability insurance, it will be a matter of interest to see how insurers classify these consultants professionally.

The State: The Value and Accountability of Expertise
In the United States, both the legislative and judicial branches of government are positioned to confer authority on ethics consultants. Although there is no state licensure of ethics consultants, state legislatures have passed other statutes related to ethics consultation. Some states provide ethics committee members with statutory immunity from litigation, and there are several state laws that promote or require the participation of ethics consultants in controversial health care decision making, such as end-of-life care.[14] These statutory actions clearly indicate that ethics consultation provides value in the eyes of legislators, or at least that there is a need to be filled.

The judicial branch has also provided some affirmation of the professional jurisdiction of ethics consultants' usefulness by allowing and utilizing expert testimony from ethics consultants. At least five landmark bioethics cases (cases creating legal precedent, conferring distinctive authority, or representing the first court appearance of a particular bioethical issue) featured expert ethics testimony (Poland, 1997; Spielman & Agich, 1999). These cases include *Wetherill v. University of Chicago*, where the court was asked to assess the qualifications of an ethics expert to testify about standards of informed consent in research; one of Dr. Jack Kevorkian's trials, in which the qualifications of renowned ethicist Arthur Caplan to provide expert testimony were challenged (Caplan, 1991); the landmark *Matter of Baby K*, regarding the futility of treatment of an anencephalic infant, which featured opposing ethics testimony from both the prosecution and the defense (see Fletcher, 1997); and a dismissed murder case against two physicians for withdrawing a feeding tube from an irreversibly comatose patient at the family's request, which was appealed and ultimately upheld in a decision drawing heavily upon bioethics expertise (Paris, 1984).

There have not yet been any successful lawsuits against ethics consultants, so their professional liability remains untested. The liability task force report notes that the *Core Competencies* report (ASBH, 1998), even if voluntary, could impact the liability of consultants by establishing a set of professional expectations, which could be presented in court as evidence of peer standards. Consequently, ethics consultants may be vulnerable not only to negligence suits, but to professional malpractice suits. Furthermore, the liability task force report soberly observes that "most consultants may not possess even a basic level" of these competencies (p. 31). Even though the *Core Competencies* report was explicitly not intended to establish a legal national standard for competence, it could nonetheless, in future cases, be construed as such a legal standard, one which many ethics consultants may not currently be able to meet. It will be a matter of interest to observe how the courts determine the legal standard of competence for ethics consultants in future cases, and the extent to which it relies on both expert ethics testimony and peer standards developed by ethics consultants.

Liability and the Jurisdiction of Ethics Consultation

The liability task force report concludes with an assessment of the advantages and disadvantages of professional liability insurance for ethics consultants. In its favor, insurance is cited as offering some protection from litigation, facilitating the clarification of the professional status of ethics consultants, and confirming appropriate assumption of liability in ethics consultation. Disadvantages of liability insurance are found to include the possible stimulation of an increase in lawsuits, and the observation that it may be in the best long-term interest of ethics consultants to clarify their professional status among themselves, rather than having it clarified by insurance companies for coverage purposes.

The liability task force report and the forgoing account of expertise and liability in ethics consultation raise questions about the coproduction of legitimacy and expertise in ethics consultation. Who decides that ethics consultants are liable for their expertise, and, who decides what that accountable expertise is? The answers to these questions depend in part, on whether lawsuits are filed against ethics consultants in the near future. Until such a time, the onus is on ethics consultants to circumscribe debates about their liability and expertise by developing some additional hallmarks of professionalism. Ethics consultants can frame these debates by clarifying their core competencies and the nature of their expertise, applying clarified core competencies as standards to establish certification and accreditation programs, and establishing a professional code of ethics.

However, the multidisciplinary field of bioethics is ambivalent about the professionalization of ethics consultation, and reluctant to demarcate jurisdictional expertise, which is partly due to external pressures to embrace the ideology of expert professionalism, rather than the social trustee professionalism in which bioethics is rooted. This ambivalence will likely provide the courts, insurance companies, and other stakeholders with considerable opportunity to define the liability and accountable expertise of ethics consultants. In response to the question, "would you like the report to indicate that liability should accrue to ethics consultations?," three members of the liability task force deferred to the courts, with one member stating, "I do not think it is our role to advise the judiciary about whether liability will accrue to ethics consultants" (ASBH, 1998, p. 43). However, three task force members also affirmed the need for professional accountability, with one member responding, "If what is meant is that ethics consultants should stop trying to avoid legal accountability, my answer is yes" (p. 43). Ethics consultants may actively address their own accountability, or may wait for the courts and insurers to establish their liability.

One manifestation of acknowledging professional accountability would be the creation of a professional code of ethics. This has been a perennial topic of debate in bioethics, the most recent round of which is published in the September/October 2005 issue of the *American Journal of Bioethics* (see main article, Baker, 2005). The debate hinges, in part, on the fact that such a code may marginalize parts of the field of bioethics, because a single professional code of ethics would be incapable of adequately addressing the various professional backgrounds and scope of work in the field. The *Core Competencies* task force report contains the rudiments of a code of ethics for consultants, although it is not presented as a code of ethics, but rather as a section entitled "Special Obligations of Ethics Consultants and Institutions" (ASBH, 1998, p. 29). The *Core Competencies* report arguably makes a strong case for developing a professional code of ethics, by noting the potential for conflict of interest and abuses of power arising from the social authority of ethics consultants.

It is worth noting that the liability task force report does not address the adoption of a code of ethics as a possible prophylaxis of malpractice lawsuits. While a code of ethics would little address the content of the advice given by consultants, considering the many medical and research liability lawsuits that have claimed conflict of interest and other kinds of professional misconduct, such a code would seem to be a prudent safeguard, clarifying to the ethics consultant, as well as to the courts and the public, what professional behavior is acceptable, and what is unacceptable. The only liability remedy discussed by the liability task force report, besides insurance, is the use of an indemnification clause in consultation contracts; the report even provides an example of one (ASBH, 2004 p. 42). However, the report also notes that one task force member opposed such indemnification, because such clauses are generally invalid in health care, and because it was deemed inappropriate for the report to help consultants evade liability (pp. 42–43).

The preceding analysis provides an account of sources of legitimacy for health care ethics consultation in the legal arena. These sources are external to the field itself, and there may well be different internal sources of legitimacy that matter primarily to ethics consultants themselves. Ethics consultants have legitimacy of technique in relation to health care professionals, who regularly seek their advice for patient-care decision making. The liability task force report, focusing on the competencies of ethics consultants, infers that technique will also be the basis of the legitimacy of ethics consultation in relation to insurance companies and the courts. Indeed, the state has affirmed the value of ethics consultant expertise, through statutory provisions for its use and immunity, and the influence of expert testimony, in the judicial forum.

However, social structure and not merely technical expertise also strongly shapes the legitimacy and liability of ethics consultation. The extent to which ethics consultants are informally perceived and formally designated as regular partners in a health care team, or as serving routine functions in health care institutions, they may share similar liability and legitimacy compared to other health care professionals with whom they work. Widespread adoption of the *ethics facilitation* approach advocated by the *Core Competencies* report (ASBH, 1998) may limit the liability of ethics consultants, by rendering their advice nondirective and non-binding. If so, ethics consultants might play a larger role in defining their own legitimate expertise by asserting competencies determined by the profession itself. If, on the other hand, consultants are named as defendants in future lawsuits, insurers and the state will take considerable interest in defining the accountable expertise of ethics consultation. In the system of professions, the perceptions of health care professionals, insurers, and the state regarding the professional status of ethics consultation (and by extension, bioethics in general) shape the legitimacy and liability of ethics consultants, by indicating their accountable expertise and jurisdictional claim.

CONCLUSION

This chapter has discussed how institutional review boards (IRBs) and health care ethics consultants may be held legally accountable for their expertise and decision making, and has argued that liability in the legal arena signifies legitimacy in the system of professions. As clients and the state come to rely increasingly on a particular expertise, and as a result confer greater authority or legitimacy upon it, the providers of that expertise can become more liable for negligence and malpractice, whereby the expert is held responsible for causing injury as a result of failing to provide the expected professional standard of care.

In the coming years, it will be a matter of interest to observe who is ultimately held responsible for the ethicality of decision making in biomedical research. Principal investigators? Their sponsors or employers? IRBs? Ethics consultants? Research participants themselves? The matter of who is held accountable has important implications for the distribution of power among stakeholders, and for the extent to which IRBs or ethics consultants may be accused reasonably of merely protecting the autonomy of powerful biomedical interests. As mentioned previously in this chapter, the court's ruling in *Grimes v. Kennedy-Krieger Institute* suggests that researchers can still be held liable for harm resulting from research, in spite of parental consent and IRB approval of research protocols. Arguably the court's decision to override the IRB's ruling in this instance was based on a relatively straightforward reading of Common Rule protections for child research participants. If the IRB had been named a defendant in the lawsuit, it would likely have been found liable, but it is not clear what impact that would have on the liability of the investigators.

It is not currently possible to examine the pursuit of state licensure as a means of strengthening the jurisdictional claims of bioethics. It is possible that licensure may never be sought for ethics consultants, if current sentiments among ethicists against certification, accreditation, and licensure are any indication.[15] However, there are other ways in which jurisdictions are shaped in the legal arena. These indicators of jurisdictional construction include the liability of emerging professional groups, admission of testimony by expert witnesses in court case hearings and decisions, and the passage of statutes promoting the use of new expertise or based on the content of that expertise.

For several reasons, the liability of IRBs strengthens, and may continue to strengthen, the jurisdictional claim of bioethics. Firstly, the legal standard of care for IRBs is derived in considerable part from federal regulations, which promulgate bioethics concepts in the requirement for IRB review of publicly funded research. Secondly, the federal regulations are ambiguous and require careful interpretation in their application to each protocol, creating a need for expert knowledge and skill by investigators, IRBs, and the courts, in order to determine the appropriate standard of care that the research enterprise must provide to study participants. This need is a jurisdictional opportunity for bioethicists, who can and do provide education and consultation to these parties, based on their knowledge of national and international ethical standards, ethical theories, and on their ethical reasoning skills. One of my case study respondents from chapter 2 predicted that research ethics will create an enormous demand for bioethicists, and consequently, research on research ethics will become a primary activity in the field. This forecast is consistent with the jurisdictional opportunity created by the increasing liability of IRBs, universities, and investigators in research litigation.

Indeed, the times do seem to be changing. A 2002 article lamenting the dearth of research ethics education in the biosciences cited the autonomy-oriented ethos of the bioscience research community, the lack of a professional bioscience ethics code and difficulty of producing one, the "absence of any traditional clients who might complain of research negligence," and the lack of disciplinary mechanisms as factors hindering the development of research ethics education (Eisen & Berry, 2002, p. 39). Three years later, a growing number of high-profile research lawsuits have presented biomedical investigators with legal allegations from participants of misconduct or injury, a real threat of disciplinary action, and an incentive to pursue research ethics training. Finally, IRB accreditation represents another way in which the liability of IRBs reinforces the bioethics jurisdictional claim. IRB accreditation, which came about through a broad-based consortium of elite research stakeholders, adds another layer of social structure to research ethics, promotes institutional isomorphism, and clarifies the expertise and performance standards required for IRB function and membership. Accreditation strengthens the bioethics jurisdictional claim by organizing and standardizing the expertise of research ethics, and affirming its utility to research institutions.

The liability of ethics consultants is even less well established than the liability of IRBs, but discourse on the liability of ethics consultants has implications for the jurisdictional claim of bioethics. Firstly, other interests in the system of professions, (including the health care professionals, who seek advice from ethics consultants; insurance companies, which assess the risk and insurability of ethics consultation; and the state, which provides some legal authorization of consultants' expertise, and will judge the liability of that expertise in future tort claims) play important roles in defining the fundamental expertise and liability of ethics consultation. The impact of other interests on the jurisdictional claims of ethics consultants is increased by the ambivalence of ethics consultation toward professionalism.

Secondly, the discourse on consultation liability reveals several sources of legitimacy that may be invoked for ethics consultation, including legitimacy of technical expertise, legitimacy of location in social structure, and legitimacy of character, or professional virtue. Expertise and social relations appear to play important roles in the external legitimacy of ethics consultation, whereas legitimacy of character does not. However, it might be argued that legitimacy of character, such as through the promulgation of a professional code of ethics, could be effectively pursued as a protection in case of malpractice lawsuits, in addition to circumscription of accountable expertise. The ASBH liability task force recognized the importance of staking the jurisdictional claims of ethics consultation carefully, as evidenced by their observations about the relative liability of different consultation approaches (ASBH, 2004). However, the extent to which ethics consultants will be able to determine their accountable expertise depends on whether they seize the opportunity to do so, as well as on the perceived utility and liability of ethics consultation, as expressed by other professional interests and the state.

Taken together, analysis of the liability of IRBs and ethics consultants illustrates the complexity of the bioethics jurisdiction. In this chapter renowned bioethicist Arthur Caplan provides a striking example of the multiple roles that characterize the work of bioethics, and that may be performed by the same person; Caplan has given expert ethics testimony in court, served as an IRB member, testified as a research ethics expert in a congressional hearing on the protection of human research participants, provided ethics consultation to gene therapy investigators, and was named as a defendant in the *Gelsinger* lawsuit. The jurisdiction of bioethics, constructed around the task of determining the right thing to do in the conduct of biomedical practice and research, encompasses experts with various disciplinary backgrounds, in multiple workplace settings, serving an array of clients, and performing a range of tasks and services.

What is the relationship of liability to forms of accountability in the other arenas where jurisdictional claims are constructed? In the case of biomedical research, the pressure for investigators to be accountable in the university workplace by producing knowledge is undoubtedly contributing to their liability in the legal arena, as plaintiffs make conflict of interest and other misconduct claims against researchers in the judicial forum. For academic bioethicists, workplace and legal accountability are also closely linked. For universities, bioethics represents, among other things, a means of tapping into opportunity structures and legitimating biomedical research; thus, it behooves academic bioethicists to provide remedies for the liability of investigators and IRBs, in order to demonstrate their own accountability to academe. As chapter 6 shall demonstrate, the public-policy recommendations contributed by the National Bioethics Advisory Commission are accountable to both the functions of research productivity assurance and research integrity assurance, and include elaborate procedural recommendations that would assure the ethical conduct of embryonic stem cell research, consequently defining the liability of stem cell researchers.

An obvious feature of bioethics jurisdiction claims in the legal arena is the codification of bioethics concepts and tools into law. The case could be made that in the legal arena, it is not the goal of bioethics to establish legal protection of their jurisdiction, but perhaps instead to shape the practice of medicine and bioscience through the codification of various bioethics concepts into law. This codification has been occurring in the federal executive branch, most notably in the Common Rule, and also in Congress and in state legislatures, which have passed laws on bioethical issues such as the privacy and confidentiality of personal health information, informed consent, and genetic nondiscrimination in employment.

As chapter 6 shall show, the impact of bioethics on federal regulations is heavily modulated not only by the expert jurisdiction of medical practice, but even more strongly by the expert jurisdiction of biomedical science, where both jurisdictions seek to protect their professional autonomy by setting limits on executive-branch oversight of science by nonscientists (see Evans, 2002). This demonstrates yet another way in which professions stake jurisdictional claims in the legal arena—in addition to protecting control of their jurisdiction through state licensing requirements, professional groups also actively seek to limit the regulation of their jurisdictions by external interests. Chapter 6 shall examine boundary work conducted by the National Bioethics Advisory Commission at the boundary between ethics and science, and the implications thereof for the regulation of biomedical research.

Endnotes

1. As Bosk (2003) noted, while a wide variety of occupations, from massage therapist to vascular surgeon, are subject to licensing and certification requirements, these requirements have different social meanings for different occupations. Bosk argued that licensure and certification are more consequential for occupations such as vascular surgery, where lay judgment of competent skill is difficult, and moreover, the consequences of incompetence are a more serious matter. IRB judgments and ethics consultations can have a major impact on the risks presented to research participants; the care of patients; and on sponsoring institutions, investigators, and care providers, should they be sued for clinical or research malpractice. Accordingly, the delineation of competent ethical expertise is important to all of these stakeholders.

2. See Lock (1996) for a comparative study on the definitions of death in North American and Japanese cultures, and Rado (1987) for an account of the network of multidisciplinary elites associated with the institutionalization of the brain-death concept.

3. The normative pressure for private-sector compliance with research oversight has two components: On the one hand, the value of public-sector IRB review is touted by policymakers and the public as a means to ensuring ethical research, and its widespread use in publicly funded research generates a certain amount of expectation that all research will comply. On the other hand, IRB review may be seen by sponsors and investigators as a means to both legitimize their research to the public, and to provide some protection against liability.

4. See Jasanoff (1998) and Halfon (1998) for analysis of the politics of expertise contained in expert testimony during the O. J. Simpson trials.

5. For an overview and categorization of problems with, and recommendations for the reform of, the current oversight system for human research, see Emanuel et al. (2004). For a substantive critique of the moral reasoning assumptions inherent in IRB reviews of research, see Eckenwiler (2001).

6. JHU's response was retrieved October 24, 2005, from http://www.hopkins-medicine.org/press/2001/JULY/010719.htm.

7. A list of accredited institutions was retrieved January 22, 2008, from http://www.aahrpp.org/www.aspx?PageID=11$1$100.

8. Universities are, of course, also taking note. For example, the University of Iowa Clinical Trials Office (CTO), established in July 1998 "to facilitate

and increase industry-sponsored clinical trials at The University of Iowa," (retrieved January 22, 2008, from http://research.uiowa.edu/cto/), has linked to its Web site a presentation entitled "Avoiding Liability in Clinical Trials" developed by attorney Kendra Dimond, who represents clients in the research and health care industries (retrieved January 22, 2008, from http://research. uiowa.edu/cto/invest-coord/forms/liability.pdf). Interestingly, at the end of 2005, the University of Iowa began requiring all of its corporate-sponsored clinical trials to be reviewed by the proprietary Western IRB, whereas clinical trials supported by other funding sources continued to be reviewed by the university's IRB. Both Western and the University of Iowa are accredited by AAHRPP.

9. This should not be construed to mean that the risks of research are greater than the risks of treatment, which can also entail improbable but grave risks (Morreim, 2004); rather, informed consent in research has a different legal basis than the basis of consent for treatment.

10. For an analysis of the positive and negative consequences of IRB liability which, on balance, finds IRB liability to be detrimental, see Hoffman and Berg (2005).

11. Negligence is not the only claim a plaintiff might make against an IRB, but it is the primary legal theory that has been tested, and is of particular interest for my consideration of the legitimacy of bioethics. For a discussion of torts of invasion of privacy and breach of confidentiality, for which IRBs could be found liable, see Hoffman and Berg (2005).

12. Morreim (2004) argued that standard tort doctrine, including the usual conceptions of negligence, battery, and informed consent, does not appropriately address conduct in the research setting, which differs significantly from medical practice in its goals, sources of duty, the nature of injuries, and the inherently subjective nature of the decision to participate in research. The efforts of Milstein and other prosecuting attorneys to introduce new claims of dignitary harm from research negligence (Dembner, 2002; Lookwood Tooher, 2005) are consistent with Morreim's observations.

13. Milstein was not the first to pursue a claim of dignitary harm. Florida Attorney Stephen Hanlon filed a class action on behalf of poor women with high-risk pregnancies who participated in a drug study. The suit contested the adequacy of the informed consent. The University of South Florida and Tampa General Hospital settled in 2000 for $3.8 million and revised the consent forms, but admitted no wrongdoing (Dembner, 2002).

14. Maryland provides statutory immunity from suits for ethics committee members, MD Code Ann., Health-Gen. I § 19–374(a) 2000. Maryland,

op. cit.; Texas, Tex. Code Ann. § 405.53(4) 2001; Montana, Mont. Code Ann. § 37-2-201(1) 1999; and Arizona, Ariz. Rev. Stat § 36–3231(B) 2000, give ethics committees statutory authority to provide advice and recommendations to medical professionals. For a more detailed discussion, see Spielman (2001) at p. 169.

15. Chapter 3 discussed the arguments of the Task Force on Standards for Bioethics Consultation against mandatory certification and accreditation (ASBH, 1998). For another account eschewing certification and licensure, see also Bosk (2003).

CHAPTER 6

BIOETHICS IN PUBLIC POLICY: THE NATIONAL BIOETHICS ADVISORY COMMISSION AND THE STEM CELL RESEARCH DEBATE

Public debate about the morality of research using embryonic stem cells is one of many biotechnology-related controversies that have arisen over the last several decades. Public bioethics advisory bodies have been a staple of U.S. public policy for addressing such societal disputes, in spite of the limited direct impact these bodies have had on science and technology policymaking. Kelly (2003) argued that public bioethics advisory bodies serve an important tacit function as boundary organizations that stabilize the border between science and politics, thus preserving the autonomy of science from incursion by other societal stakeholders. These boundary organizations succeed in bounding and controlling the controversy by constraining the set of issues and viewpoints that are addressed, and by dictating the decision-making strategy in ways that privilege participation of some stakeholders over others and veil the intensity of the controversy.

Science is interdependent with the state, which exchanges resources for the conduct of science in return for the economic and technological fruits of scrupulous scientific activity. This relationship can be interpreted as a contractual one between government principals and researcher agents that is mediated by boundary organizations which facilitate the assurance of mutual goals (Guston, 2000) and stabilize the science-politics boundary by negotiating the boundary's contingencies within the organization. Boundary organizations employ specialized mediators, who serve as dual agents to both the government principals and the scientist agents, catalyzing collaboration between the state and science.

The National Bioethics Advisory Commission (NBAC), a presidential advisory body created during the Clinton Administration, constituted such a boundary organization. The creation of NBAC both directly legitimated bioethics, and provided opportunities for the further institutionalization of bioethics in the public-policy arena. In addition to mediating the tensions between politics and science, NBAC was also faced with negotiating the boundaries between science and ethics on the one hand, and between ethics and public policy on the other.

This chapter examines the boundary work conducted by the U.S. National Bioethics Advisory Commission (NBAC) at the borders between science and ethics, and between ethics and public policy, in NBAC's deliberations and recommendations on embryonic stem cell research. Specifically, it examines the coupling of scientific and ethical uncertainty, and the coupling of research productivity and integrity assurance, in NBAC's deliberations on embryonic stem cell research, and discusses how NBAC's boundary work ultimately served to reinforce the authority of science and to marginalize conflicting civic-sector concerns.

This chapter begins by sketching the historical context of the relationship between philosophers and public policy in the United States, which reveals pervasive tensions that underpin the engagement of bioethics with public policy embodied by NBAC. It then reviews different perspectives on the functions of executive advisory bodies generally, and related criteria for their success. Next, it provides some background on NBAC and the embryonic stem cell controversy, and describes features that characterize NBAC as a boundary organization. The heart of the chapter describes NBAC's boundary work between science and ethics, and between ethics and public policy; this chapter examines how science and ethics are coupled in the construction of the embryo as a boundary object, how the boundary between ethics and public policy is negotiated in deliberations about the role of federal funding and oversight in embryonic stem cell research, and how NBAC's recommendations serve the mutual goals of science and the state, while at the same time legitimating bioethics.

PHILOSOPHERS AND PUBLIC POLICY

American philosophers' involvement in public policy and debate has changed considerably over the course of the 20th century. As Albert Jonsen (1998a) explained,

> The golden era of American philosophy spanned the last quarter of the nineteenth century and the first quarter of the twentieth. William James and Josiah Royce at Harvard, and John Dewey and George Herbert Mead at the University of Chicago, cultivated a style of philosophizing that combined an erudite appreciation of the classic philosophers with an appealing public voice about contemporary issues...The tradition of the public philosopher culminated with these men, who incessantly addressed the public about the life of the nation and its affairs. Yet, even as these public philosophers were thriving, philosophy was becoming more technical, more academic, and more reticent about public affairs. (p. 68)

The tradition of the public philosophers faded as Continental logical positivism from the Vienna Circle came into vogue, bringing its devastating critique of normative ethics. Jonsen (1998a) described a major shift occurring in moral philosophy, namely that metaethics (a term coined in 1949), the study of the meaning of ethical terms and concepts, came to supersede normative ethics, which is concerned with discerning right and wrong, in philosophical discourse. In the middle of the 20th century, it was political philosophers like Hannah Arendt and Sidney Hook, rather than moral philosophers, who tackled some of the moral challenges of the day, including the Holocaust, the Nuremberg trials, nuclear weapons, and McCarthyism. Finally, the political climate of the 1960s, culminating in Vietnam War unrest, compelled philosophers to address society's problems again, at the urging of linguist Noam Chomsky (Jonsen, 1998a).

However, as philosophers were once again called to serve in the public-policy arena, they soon experienced firsthand the fundamental tensions between philosophy and public policy. In essence, philosophy is fundamentally aimed at determining *truth* through analytical reasoning, while public policy is concerned with optimizing the *consequences* of state action through political deliberation (Brock, 1987). Furthermore, philosophers, perceived as the quintessential ivory-tower academics, must demonstrate their credibility in the pragmatic realm of policy. Weisbard (1987) was quite pessimistic about the capacity of philosophy to add value to public policy, contending that philosophical scrutiny often demands standards for justification that few realistic policy proposals can meet.

The relationship between ethics and politics is further complicated by fundamental, conflicting convictions within the American ethos.[1] On the one hand, our classical liberal tradition champions a secular, minimalist morality, consistent with the protection of individual freedom and equality, and tolerance of pluralistic differences. On the other hand, our American culture, tracing back to its Puritan roots, has had a continual penchant for moralism, in that our ethos "is strongly tempted to endow various aspects of life with moral meaning in a capricious way" (Jonsen, 1998a, p. 391). Jonsen noted that in the United States, relatively mundane matters such as diet, exercise, and smoking have taken on "moral" dimensions, often inspiring moral crusades.

The competing liberal and moral impulses of the American ethos are evident in the abortion debate, which reveals the more general problem of how to resolve moral disagreements in a pluralistic society. Arguments supporting the legality of abortion invoke the right of women to make their own choices and control their own bodies, whereas arguments supporting the criminalization of abortion attribute to fetuses the status of personhood and the accompanying rights, and cite the wrongness of killing. Few issues in the American political landscape are as polarizing. The prospect for reaching consensus on abortion, or on any similarly contested issues, is made all the more doubtful by the increasingly politicized climate that has taken hold in the United States since the 1970s.

It now appears that the tension between philosophy and public policy has manifested in the disjuncture between philosophy and bioethics, as described in chapter 3 by a few of my case study respondents. Academic philosophy is once again averse to "getting its hands dirty" in the policy world, whereas bioethics is accused of being too driven by current events in the media, in the quest of bioethics to influence policy. As this chapter will show, bioethics, as proffered by the National Bioethics Advisory Commission, adopts a public-policy approach rather than a more philosophical approach, such that political consequences ranked in importance above moral truth, in spite of the fact that the latter were arrived at by analytical reasoning.

Bioethics has indeed become a fixture in the federal-policy landscape, facilitated by the current science policy paradigm of collaborative assurance, which succeeded the relatively laissez faire, social-contract science policy approach of the post-World War II era (Guston, 2000). The institution of science, and its social-contract relations with the military, came under fire in the social protests of the 1960s, which deemed science no longer trustworthy to manage its own integrity. In the subsequent collaborative assurance regime, government and science worked together through a variety of jointly operated mechanisms to mutually, collaboratively assure the productivity and integrity of science, through the leverage of state-supplied resources.

THE FUNCTIONS OF EXECUTIVE ADVISORY BODIES

Presidential commissions are perhaps the most recognizable of the executive advisory bodies in the United States. The political science literature on presidential commissions has identified two broad categories of organizational purpose, problem solving and conflict management (Sulzner, 1974).[2] Problem solving includes defining and investigating a problem and providing recommendations for its solution. Presidential commissions assist presidents in making informed decisions, arguably in the face of factual and political uncertainty (Wolanin, 1975) or inadequacies in the existing executive advisory mechanisms (Flitner, 1986). Conflict management includes strategies such as building consensus, pacifying political opponents, and encouraging public support for a course of action that policy makers have already determined (Bell, 1969; Cleveland, 1964). Presidential commissions also manage conflict through symbolic functions, such as expressing the current administration's concern about a particular problem, promoting societal awareness of a problem (Flitner, 1986), serving as a lightning rod for controversial issues, or delaying presidential action until a more propitious time. Delay can be viewed either negatively, as constituting an evasion tactic (Cleveland, 1964), or positively, as creating a constructive cooling-off period in policy making (Sulzner, 1974).

Presidential commissions have frequently been disparaged as ineffective, as evidenced by articles with such titles as, "Why Commissions Don't Work" (Schmitt, 1989), "The Ambiguous Legacy of American Presidential Commissions" (Graham, 1985), and "Nuclear Committee Plays it Straight—And Draws Criticism from All Quarters" (Lanouette, 1981). Some of the criticism of commissions stems from disagreement over the criteria for their success, which of course vary with their variously ascribed functions. In reference to bioethics advisory bodies specifically, the Institute of Medicine's Committee on the Social and Ethical

Impacts of Developments in Biomedicine identified three categories of criteria for success: intellectual integrity, sensitivity to democratic values, and effectiveness, presumably signified by the implementation of a body's recommendations (Bulger et al., 1995). Bradford Gray (1995) further noted that the definition of success depends on the time horizon of evaluation (i.e., short-term vs. long-term impact), and the scope and intended audience of reports from advisory bodies.

There is some concurrence about the success and failure of federal bioethics bodies (for a review, see OTA, 1993); two former federal bioethics commissions in particular are recognized as having been influential in protecting human research subjects and in expanding the autonomy of patients receiving medical care: the National Commission for the Protection of Human Subjects of Biomedical and Behavioral Research (National Commission) and the President's Commission for the Study of Ethical Problems in Medicine and Biomedical and Behavioral Research (President's Commission). In response to the Tuskegee scandal, the National Commission produced the landmark Belmont Report in 1979, which laid out the basic principles to guide the ethical conduct of research with human subjects, and enumerated the functions of institutional review boards to assure the application of those ethical principles. During the early 1980s, the President's Commission issued several reports on a variety of issues, most notably *Deciding to Forego Life-Sustaining Treatment* (1983). These reports have been widely cited in court cases and medical ethics education (see Bulger et al., 1995; McAllen & Delgado, 1984). In these two examples, "success" can be inferred to mean influence, or contribution to the reform of medical and scientific practice and regulation, particularly over the long term.

Science and technology studies (STS) perspectives, including the concept of the boundary organization, provide a more nuanced understanding of the "success" of advisory bodies, the role of science policy advisors generally, and the ways in which stakeholder interests shape advisory-body results, function, and strategies for managing political conflict. The boundary work approach exposes the ways in which particular cultural maps serve the interests of social actors who contend for authority, power, and resources. Gieryn (1983) defined boundary work as "the attribution of selected characteristics to the institution of science (i.e., to its practitioners, methods, stock of knowledge, values, and work organization) for purposes of constructing a social boundary that distinguishes some intellectual activity as 'non-science'" (p. 782). Scientists, like other professionals, perform boundary work to distinguish themselves from other enterprises as they compete for crucial authority, power, and resources.

Boundary work is not a one-off event, but rather an ongoing activity shaped by the "structural contexts of available resources, historical precedents, and routinized expectations that enable and constrain the contents of a map and its perceived utility or accuracy in the eyes of users" (Gieryn, 1999, p. 10). Thus, it is essential to apprehend the *context* of boundary work in order to discern both the *goals* and the *strategies* of the boundary work performed by a particular social group. Boundary work analysis provides the beginnings of a cultural account of the changing allocations of power and resources among social actors over time, and invites us to reflect on alternative cultural maps that could be drawn.

STS scholars have described several ways in which boundary work is performed at the border of science and politics. Jasanoff's (1990) study of science advisory committees found that policy making is facilitated by the intentional blurring of the boundary between science and policy, and conversely, rendered more difficult when scientific advisors and policy makers attempt to strengthen distinctions between science and politics. Guston (2000) described how boundary organizations facilitate collaboration across the science-politics border in order to achieve the mutually desired goals of assuring the productivity and integrity of science. To do so, boundary organizations employ specialized mediators who serve as dual agents to both government and scientific principals, catalyzing collaboration between the state and science, and stabilizing the science-politics boundary by negotiating the boundary's contingencies within the confines of the organization. Kelly (2003) argued that bioethics advisory bodies, as boundary organizations, contain and stabilize the tensions which exist between science and politics, thereby protecting science and policy practices from the scrutiny and interference of other stakeholders. Such protection is afforded in part by the use of an overlapping consensus approach to bioethics, in which academic reflection on shared societal values is used to determine the ends of public policy (see Rawls, 1987). A consensus approach can result in the masking of underlying philosophical issues, the underestimation of risks and opposition, neglect of unpopular views, and the exclusion of some alternatives and information from consideration (Bulger et al., 1995), and in the case of public bioethics bodies, consensus simultaneously privileges bioethics experts as moral arbiters and limits and mediates the participation of other interests in the deliberations (Kelly, 2003).

Evans (2002) described how the selection of a particular form of argumentation resulted in the substantive thinning of the public bioethical debate over human genetic engineering, in the proceedings of the President's Commission. Different styles of argumentation, not surprisingly, produce different policy results. *Substantive rationality* in public-policy discourse encourages debate about multiple policy options (such as research with both embryonic and adult stem cells, and with other biotechnologies), long-term effects (of pursuing those approaches), and the prohibition of incongruous approaches. In contrast, *formal rationality* focuses public-policy discourse on one discrete policy option (e.g., embryonic stem cell research) and its immediate effects, and favors limited exploratory development of that one approach over a prohibition of it (in order to better determine policy consequences, and to observe how the issue evolves over time). In essence, the thinning of bioethical debate is part of the larger phenomenon of the Weberian rationalization of society. Evans finds that the thinning of the debate by the commission excluded the option of not engaging in any human genetic engineering. More generally, Evans (2002) observed that such thinned debate favors some interests over others, and shifted the locus of debate away from the public and toward the bureaucratic, technocratic state.

The interests of different stakeholders, including the professionals who staff commissions, clearly shape the operation of advisory bodies and the trajectory of their inquiries. Rather than asking whether a particular commission or report has been a success by virtue of having its recommendations implemented, STS perspectives encourage us to ask whose criteria for success shape a commission's operation and products, how those criteria become legitimate, and with what consequences for all who hold a stake in the public good of science. The next section of this chapter provides some context for the analysis of the coupling of science and ethics resulting from the boundary work performed the National Bioethics Advisory Commission in its deliberations on stem cell research.

NBAC AND THE EMBRYONIC STEM CELL RESEARCH CONTROVERSY

Several standing federal bioethics bodies went before NBAC, beginning with the National Commission for the Protection of Human Subjects of Biomedical and Behavioral Research, which operated from 1974 to 1978. However, more than a decade passed between the termination of the President's Commission in 1983 and the creation of NBAC in 1995.[3]

The direct origins of NBAC can be traced to 1992, when the Senate debate on reauthorization for the National Institutes of Health restimulated congressional interest in a formal role for bioethics in federal governance, 9 years after the termination of the President's Commission. Senators Mark Hatfield, Edward Kennedy, and Dennis DeConcini asked the Office of Technology Assessment to review the history of bioethics in federal policy, and to evaluate prior bioethical policy approaches, in order to help Congress develop potential strategies for examining policy issues with biomedical and ethical import. Senators Hatfield and Kennedy introduced a bill in the first session of the 103rd Congress (1993) to establish a national bioethics commission.

That same year, the Department of Energy (DOE) and the National Institutes of Health (NIH) appealed to the White House Office of Science and Technology Policy (OSTP) for the establishment of a standing national bioethics commission. NBAC's charter was signed in July

1996 by Jack Gibbons, then assistant to the president for Science and Technology Policy. NBAC's first priorities were to address "protection of the rights and welfare of human research subjects; and issues in the management and use of genetic information" (Executive Order 12975 of October 3, 1995, Sec. 5a). In response to the report issued by the Advisory Committee on Human Radiation Experiments in 1995, President Clinton also required relevant executive-branch agencies to review their human subjects protection policies, and to report their results to NBAC. Furthermore, NBAC was granted the authority to deliberate on additional issues raised by the general public, other federal bodies and organizations, or NBAC itself.

Although NBAC began with vocal support from both the executive and legislative branches, the congressional pledge to balance the federal budget in 1996 provided little funding commitment to the operation of NBAC. For the first 2 years of its existence, NBAC relied upon volunteered funding collected from various agencies by the efforts of Senator Mark Hatfield, a strong advocate for NBAC and the then chair of the Senate Appropriations Committee. Even with such a powerful champion, NBAC's initial funding was less than that of previous federal bioethics bodies. On National Public Radio, bioethicist Arthur Caplan quipped that Spartan funding made NBAC more of a Yugo than a Cadillac (citation from National Public Radio, October 4, 1996). By fiscal year 1999, NBAC's budget of $2 million, a increase of 5% from the previous year, came entirely from the Department of Health and Human Services (DHHS), as per the executive order and the charter establishing NBAC. These founding documents also required DHHS to provide "management and support services" for NBAC, raising the estimated annual cost for NBAC operations to $3 million in 1999 (NBAC, 2000, p. 28).

Initial plans for the routine operation of NBAC changed considerably as NBAC was asked to address an increasing number of topics. After the first meeting, two subcommittees were formed to address the two priorities mandated in the charter, namely, the protection of human research subjects and the management and use of genetic information. Chairman Harold Shapiro had anticipated quarterly meetings at the NBAC's outset, but in its first year of operation, the full commission met eight times, the human subjects subcommittee met six times, and the genetics subcommittee met five times. NBAC's workload increased unexpectedly in February 1997 with the announcement of Dolly the cloned sheep, when the president asked NBAC to address human cloning. President Clinton's request for a report on stem cell research in November 1998 was another unanticipated request, and during the 10 months that NBAC worked on the stem cell report, they were concurrently working on other projects, sometimes three at once.

In November 1998, three separate reports of the isolation and culture of human and hybrid embryonic stem cells (Shamblott et al., 1998; Thomson et al., 1998; Wade, 1998) prompted President Clinton to assign two tasks to NBAC: (1) NBAC was asked to promptly examine the implications of Advanced Cell Technology's announcement that they had created hybrid human-cow stem cells and to report back to the president as quickly as possible, and (2) President Clinton requested that NBAC "undertake a thorough review of the issues associated with such human stem cell research, balancing all ethical and medical considerations" (NBAC 1999a, vol. 1, p. 89).

Kelly (1994) notes that one recurrent theme in the history of federal bioethics advisory bodies has been the perpetual struggle between legislators and mission agencies to control jurisdiction over the approval of biomedical research funding, with the result that ethics investigations of bioethics bodies are directed toward political power and resource control. Both the DHEW Ethics Advisory Board's 1979 report (EAB, 1979/1988) on in vitro fertilization research and the HFTTR panel's 1988 report (HFTTR Panel, 1988) were instigated by grant applications requesting government funding for controversial research. The embryonic stem cell research controversy also grew out of concerns about government funding of the research. The reports of stem cell research breakthroughs immediately provoked reaction at the National Institutes of Health (NIH), where concerns quickly arose that federal law might well prohibit public funding of research using the new stem cells. The statute in question was a rider attached to the DHHS appropriation in

1996 and subsequent years, which has barred the use of federal funds for any activities involving either the creation of human embryos for research purposes, or research in which human embryos are destroyed or subjected to risk of injury greater than that allowed for research on fetuses in utero by the Common Rule (45 CFR 46) and the Public Health Service Act (42 USC 289).

Harold Varmus, the then NIH director, requested an opinion from DHHS about the legality of using DHHS funds to support stem cell research. Varmus chose the occasion of his testimony at NBAC's January 1999 meeting to announce the DHHS decision: DHHS general counsel Harriet Rabb ruled that research on the use, although not the derivation, of stem cells from embryos could receive federal funding, explaining that "statutory prohibition on the use of funds appropriated to HHS for human embryo research will not apply to research utilizing human pluripotent stem cells because such cells are not a human embryo within statutory definition" (NBAC, 1999b, pp. 17–18). In February 1999, 70 members of Congress responded to Rabb's ruling by signing a letter urging DHHS to reverse Varmus's decision to allow NIH funding of embryonic stem cell research. In turn, 73 prominent scientists endorsed a letter pressing the administration to support Varmus's decision (see Lanza et al., 1999), and 33 Nobel laureates, representing the American Society for Cell Biology, signed and sent another letter directly to President Clinton and Congress, supporting stem cell research.

Against this political backdrop, NBAC developed a report and recommendations for human stem cell research. The boundary work performed by NBAC is only one of several instances of boundary work performed by various interests in the stem cell controversy. Consider, for example, the proceedings of the Ethics Advisory Board convened by Geron Corporation, which funded the research behind two of the three stem cell research breakthroughs announced in November 1998 (Hastings Center, 1999) and other stakeholders who advocated for a research focus on stem cells isolated from adult tissue (see Vogel, 2001). NBAC did not draw the definitive cultural maps of science, ethics, and politics in the debate. However, NBAC represents the administration's efforts to incorporate bioethics into public policy at the federal level, and a substantial collection of stakeholder organizations were interested in the process and results of NBAC's analysis (Eiseman, 2003).

NBAC AS A BOUNDARY ORGANIZATION

Boundary organizations internalize the contingent character of the science-politics boundary, and carefully negotiate boundary contingencies by creating and using *boundary objects* and *standardized packages* to mediate collaboration between the interests of scientific and political stakeholders. Boundary objects are socially constructed entities that are "adaptable to different viewpoints and robust enough to maintain identity across them" (Star & Griesmer, 1989, p. 387), such that different stakeholders may refer to the same object in shared discourse, but attach different, viewpoint-specific meanings and ends to it. An embryonic stem cell is such a boundary object, just as NBAC's published report (on ethical issues in research with stem cells) is also a boundary object. Standardized packages combine boundary objects with consistent rules, procedures, and norms, which are sturdy enough to change local practices across settings in which they are adopted (or more likely, externally imposed) and applied. Contemporary ethical review of research with human participants employs several of these standardized packages, including institutional review boards, and the federal regulations and ethical principles these boards apply in evaluating proposed research.

The current section of this chapter examines how NBAC manages the coupling of scientific and ethical concepts, consequently managing the boundary between science and politics. It describes the ways in which NBAC manifests the three criteria of boundary organizations outlined by Guston (2000), paying particular attention to the ways in which NBAC employed the boundary objects of human cells and embryos, and the standardized packages—that is, science policy tools—that have been developed over the last few decades to manage biomedical research in the name of public good. The analysis employs discourse analysis of the meeting transcripts and final report of NBAC that deal with the ethics of human stem cell research, as well as NBAC's charter (see appendix for additional methodological details). This section argues that NBAC's boundary work (particularly the ways in which science and ethics, and ethics and public policy, are coupled) served ultimately to reinforce the authority of science and marginalize conflicting civic-sector concerns.

Criterion 1: A Boundary Organization Exists on the Frontier of Relatively Distinct Social Worlds, With Distinct Lines of Responsibility and Accountability to Each

As a presidential advisory commission, NBAC had automatic responsibility and accountability to President Clinton. However, as discussed earlier, the creation of NBAC was proposed and supported by several members of Congress and executive-branch agencies; by fiscal year 1999, NBAC's $2 million budget came entirely from the purse of the Department of Health and Human Services. NBAC's charter required it to provide an annual report to the President's National Science and Technology Council and to appropriate congressional committees, which was to include summaries of NBAC activities and recommendations, as well as responses to those recommendations received from government departments and agencies, and other entities. NBAC's formal ties to the executive and legislative branches were further reinforced by congressional requests for testimony from NBAC staff and commissioners, and by NBAC's regular use of expert testimony from an array of executive-branch agency officials.

NBAC's accountability to the world of science was built into its charter, which required NBAC's membership to be "approximately evenly balanced between scientists and nonscientists" (NBAC, 2000, p. 31). The organizational charter specified that NBAC consist of 18 presidentially appointed, nongovernment members, drawing at least 1 expert from each of the fields of philosophy/theology, social/behavioral science, law, medicine/allied health professions, and biological research. NBAC was additionally required to possess at least 3 members from the general public, bringing other expertise; to roughly balance the number of scientists and nonscientists; and to seek equitable geographical, ethnic, and gender representation. Of the 17 NBAC commissioners who served at the time of the stem cell research deliberations, 7 were women, and at least 3 represented ethnic minorities. Commissioners included 2 PhD biologists, 5 MDs, 1 RN, 3 jurists, 3 PhD philosophers or theologians, 2 PhD social/behavioral scientists, 7 members of the Institute of Medicine, and 3 representatives of the general public, including the executive director of a mental health advocacy organization, and a representative from the biotechnology industry (these degrees and categories are not mutually exclusive). However, only 15 commissioners deliberated on stem cell research, because 2 commissioners eventually recused themselves from the stem cell deliberations.[4] Of the remaining 15, 5 had prior experience serving on federal bioethics bodies or in other federal organizations.

Although NBAC members were somewhat diverse with respect to gender, ethnicity, and geographical distribution, critics found them to be ideologically biased; one Catholic commentator found NBAC to be "stacked with abortion, euthanasia and eugenics supporters" (Meehan, 1996),[5] citing numerous links to prochoice organizations.[6] As per the recommendation of the Advisory Committee on Human Radiation Experiments, NBAC's first priorities were the protection of human research subjects, as well as the management and use of genetic information. As a result, two advocacy groups for radiation experiment subjects sought the appointment of a research abuse victim to NBAC. The White House Office of Science and Technology Policy declined the request, allegedly due to concerns about the ability of victims to be objective (Meehan, 1996).

While Meehan (1996) justifiably notes that there were several current and former human subject *researchers* appointed to NBAC, it should also be pointed out that research subjects are rightfully represented in NBAC. Commissioner Flynn was the executive director of the National Alliance for the Mentally Ill, which both sponsors research and has many members who are research subjects; Flynn also has a daughter with a severe psychiatric disorder who has participated in clinical research trials. Unlike the radiation victim advocacy groups, the American Psychological Association, the Biotechnology Industry Organization, and Senators Mark Hatfield (the Republican, Senator for Oregon) and Daniel Moynihan (the Democrat Senator for New York) were all successful in having nominees appointed to NBAC.

The NBAC's accountability to science was also arguably reinforced by the commissioners' strong identification with academic culture (11 of the commissioners were university professors or administrators), as well as their reliance on scientific expertise to inform their deliberations.[7]

While NBAC's accountability to political and scientific principals was manifested formally in its contractual relationships to policy makers and scientific professionals, and its dependency on those principals for funding and critical information, NBAC's accountability was also informally evident in its deliberations, particularly in the careful attention commissioners gave to the practical concerns of both bureaucrats and scientists, and in designing workable oversight mechanisms, as shall be described shortly.

Criterion 2: A Boundary Organization Involves the Participation of Both Principals and Agents, as Well as Specialized Mediators

The role of politicians and bureaucrats as principals of NBAC is fairly straightforward. President Clinton requested that NBAC "undertake a thorough review of the issues associated with such stem cell research, balancing all ethical and medical considerations" (NBAC, 1999a, vol. 1, p. 89). Congress was also very interested in NBAC's analysis, as were NIH and FDA bureaucrats, given these parties' competing interests in the federal control of both funding approval and research oversight. As principals of NBAC, the scientific community has primarily pragmatic concerns about ethics; most researchers steer clear of publicly debating the ethical implications and propriety of stem cell research, but are very interested in having policy makers spell out clear ethical guidance for the practice of research.

The pragmatic interest of stem cell investigators was reflected in the testimony given to NBAC by researchers James Thomson and John

Gearhart, whose work on stem cell derivation had instigated NBAC's investigation. At the University of Wisconsin, Thomson had isolated embryonic stem cells from leftover embryos donated by couples who had undergone infertility treatment, and Gearhart had isolated embryonic germ cells from electively aborted fetuses at Johns Hopkins University.[8] Although these investigators supported their research entirely with private funds from the Geron Corporation (see Marshall, 1998), both of them experienced similar difficulties navigating university requirements (based on federal regulations) for the ethical oversight of their research. In his testimony, Gearhart recounted,

> Now we ran into our first series—I won't refer to them as obstacles, but certainly of review. It has taken us actually a several-year period to put into place many of the requirements that were necessary to pursue the work. I wish that we had had at our disposal a committee, like Dr. Varmus [does], who could give us a determination on this in a period of a few months, perhaps. (NBAC, 1999b, p. 36)

When asked to comment on the oversight function of NIH, Thomson answered,

> from an investigator's point of view, we just want a set of rules for what's appropriate and not appropriate, what's ethical and what isn't ethical…You can do basically what you want to do at a university but I don't think that's appropriate. (NBAC,1999b, p. 64)

While they recognize the need for research oversight, and take seriously their responsibility to conduct research ethically, these investigators ultimately want to be provided with clear rules and norms, and efficient oversight, so that they can get on with what they see as the practice of science.

NBAC commissioners and staff served as specialized mediators for the political and scientific principals concerned with embryonic stem cell research. As a dual agent to both principals, NBAC sought to design flexible oversight for unpredictable science. In addition to uncertainty about exactly what therapeutic promise stem cell research could fulfill, and how soon, there was additional uncertainty regarding the promise of stem cells derived from adult tissue sources versus stem cells derived from controversial embryonic sources. After concluding that the therapeutic promise of embryonic stem cells was too compelling to pass up, NBAC deliberated on which embryonic stem cell sources could be appropriate for use in federally funded research. As agents for their political principals, NBAC conducted fact-finding and ethical and technical analysis, identified a moderate public-policy position, and designed oversight mechanisms to assure integrity; as agents for scientific principals, NBAC sought to minimize hindrance to scientific practice and progress, provided ethical guidance and procedures, and produced indicators of scientific promise and integrity. NBAC also reinforced the authority of science by emphasizing the importance of scientific state-of-the-art science and deferring to scientific expertise to set research agendas and priorities, through the merit assessments of peer review. It is worth noting that NBAC could have chosen to provide advice on the agenda of stem cell research on the ethical grounds of the just distribution of resources; instead, NBAC chose to recommend only which sources of embryonic stem cells were appropriate for use in federally funded research, relying more on the testimony and analysis of professional ethicists than on theological views of ethical science.

Criterion 3: A Boundary Organization Provides a Space Legitimating the Creation and Use of Boundary Objects and Standardized Packages

NBAC, like the bioethics advisory bodies that preceded it, has provided a space for identifying and fine-tuning the boundary objects and standardized packages that have been developed by the interdisciplinary field of bioethics. These boundary objects and standardized packages are the focus of discussion in the next section of this chapter. The central boundary object constructed by NBAC is the embryo, a potential source of stem cells. As shall be demonstrated, NBAC's boundary work exhibits a coupling of the moral status of the embryo to current scientific understanding of cellular development and differentiation. The standard-ized packages, or funding and oversight structures, recommended by NBAC served the mutual interests of the political and scientific prin-cipals, and reinforced the position of bioethics as a legitimate arbiter of ethical research practice, while marginalizing theological perspectives and narrowing the range of ethical issues addressed in NBAC's report and recommendations. Then, this chapter will examine boundary work performed in the construction of the embryo as a boundary object, the expanding role of federal funding and oversight, and the embryonic stem cell research oversight approaches recommended by NBAC.

THE EMBRYO AS BOUNDARY OBJECT

Developments in somatic cell nuclear transfer (SCNT), or cloning, and in stem cell science at the end of the 20th century increased the flexibil-ity of human embryos as boundary objects at both the science-politics boundary, and the science-ethics boundary. The birth of the cloned sheep Dolly in 1997 demonstrated that the nucleus of a terminally differentiated cell—that is, a cell whose genetic expression has been specialized for a particular adult function—can be deprogrammed to have the totipotency of a fertilized egg, or zygote. "Totipotency" refers to the zygote's ability to divide repeatedly and differentiate permanently into any specialized cell type found in the adult organism.

Similarly, advances in stem cell research reported in November 1998 by the laboratories of James Thomson and John Gearhart demonstrated that stem cells from human embryonic and fetal sources could be isolated and cultured.[9] These stem cells are *pluripotent*, meaning that they can proliferate and differentiate into several, but, unlike totipotent cells not all of the specialized cell types found in an adult organism.[10] In adult organisms, some stem cells remain, which retain the ability to divide and differentiate into a limited variety of cell types to renew tissues through-out the life of the organism. Hence, isolated embryonic stem cells have considerable scientific and therapeutic promise in the generation of new tissue for the treatment of injury and disease.

With these breakthroughs in cloning and stem cell science, investigators were coming closer to being able to dedifferentiate cells backwards through their developmental programs, and to directing the forward differentiation of pluripotent cells into targeted cell types.[11] As a result, the perceived identities of cells are becoming more plastic, and different cell types—including fertilized eggs—are more easily viewed as constituting a fluid continuum of cell types, rather than discrete categories. The enhanced plasticity of cell identity, or alternatively, the increased uncertainty of cell identity, enters the ethical debate on the status of embryos with respect to personhood.

For those who believe that personhood begins not at the instant of fertilization, but at some point later in development, the enhanced plasticity of cells and their genetic programming reinforces treatment of zygotes and similar cells (such as SCNT products) as being little or no more deserving of special respect or treatment than other cells in biomedical research. That is, zygotes are viewed as no different in status from other cells; they just have developmentally different, but malleable, genetic programs. However, for those who believe that human personhood begins at fertilization, substantial caution is warranted to treat any potentially totipotent cell as a potential human life. Furthermore, in this viewpoint, the only way to determine whether a cell is totipotent is to implant the cell in a woman's uterus and let it develop, which would be unethical because the resulting being might be irreparably damaged by the manipulations that produced it. Therefore, any cell that might be totipotent ought to be treated with the utmost respect and care.

The embryo's flexible identity as a boundary object extends beyond the aforementioned concerns about its potential status as a human life. For disease advocacy organizations and patients, such as those concerned with Parkinson's disease, embryonic stem cells represent the promise of treatments or cures for numerous chronic and life-threatening diseases. Daniel Perry, executive director of the nonprofit organization, Alliance for Aging Research, provided expert testimony to NBAC advocating research on stem cell lines derived from embryos, arguing that "this research is too momentous, too large in its potential benefits, to impede, to stop, or to slow the thrust of current scientific inquiry" (NBAC, 1999b, p. 68). For the higher education sector, the embryo represents a significant opportunity structure for research revenues, from both private and public sector sources. In March 2001, the presidents of the American Council on Education, the Association of American Universities, and the National Association of State Universities and Land-Grant Colleges, together with 112 university presidents and chancellors, wrote to DHHS Secretary Tommy Thompson, supporting federal funding for embryonic stem cell research (Eiseman, 2003). Analyzing the embryo as a boundary object enables us to illuminate the complex and multiple meanings of a contested object, and as Fujimura (1992) explains, the boundary object concept "promotes our understanding of translation efforts in the management of collective work across worlds," (p. 175) the unenviable task with which NBAC was faced.

NBAC was fully aware of the ways in which new developments in stem cell science were increasing the flexibility of the embryo as a boundary object in the debate over embryonic stem cell research. Overtly, the breakthroughs in the ability to isolate and culture embryonic stem cells brought science closer to making stem cell therapies a reality. This heightened therapeutic potential compelled President Clinton to request from NBAC "a thorough review of the issues associated with such human stem cell research, balancing all ethical and medical considerations" (President's letter to Chairman Harold Shapiro of NBAC, 11/14/98; reproduced in NBAC, 1999a, vol.1, p. 89). In the end, the promise of new stem cell therapies tipped NBAC's scales in favor of limited federal sponsorship of human embryonic stem cell research: "we have found substantial agreement among individuals with diverse perspectives that although the human embryo and fetus deserve respects as forms of human life, the scientific and clinical benefits of stem cell research should not be foregone" (NBAC, 1999a, vol. 1, p. 11).

Furthermore, over the course of their deliberations, several of the commissioners also acknowledged that scientific information often shapes our normative perceptions, despite scientists' allegations that science cannot answer questions of ethics. At NBAC's February 1999 meeting, commissioner and geneticist David Cox asserted, "there is no scientific data that is going to answer that question [of when life starts]...It is not a scientific question now, despite the fact that people try to make it such" (NBAC, 1999c, pp. 102–103). However, at the next day's meeting, Commissioner Brito, an academic pediatrician, observed,

> It seems to me that the key here is going to be to *emphasize the scientific advances that have come about and how they may have changed our perceptions, our ethical viewpoint* [italics added]. Even though David's not here to discuss this today, yesterday he said that science is irrelevant in terms of some of the ethical issues. I disagree with that. The more I've thought about that, I think the science has made it [*sic*] more relevant in terms of how we look at embryonic development because of the new findings. And I think the key here is going to be to highlight historically, in recent history, really, the last 30 to 40 years, of *how science has advanced to the point where we now understand embryonic development better, and how that may change some of the ethical viewpoints and public-policy viewpoints on embryos* [italics added]. (NBAC, 1999d, p. 64)

Commissioner Cassell, another physician, agreed, while Commissioners Shapiro and Capron noted that Brito's assertion was consistent with President Clinton's request, which was motivated at the outset by new scientific developments.

In essence, I am arguing that the president and the commissioners recognized (and uncritically accepted) that embryological science and moral understanding are constructed in tandem, or are coupled to one another. While researchers have come to recognize that invoking scientific explanations to justify moral claims serves to delegitimate and politicize science, especially in the case of human embryos' moral status, the president and some of the commissioners acknowledged that science shapes our reality and beliefs, but failed to reflect critically on the normative influence of science.

Acknowledging that scientific understanding moulds our moral perceptions implies that moral standards will evolve in tandem with perpetual scientific progress. As Commissioner Holtzman explained,

> An embryo is what it's always been—namely, something that given under normal circumstances, ordinary circumstances, goes on to become a kid. *What's changed is what is within the realm of the ordinary and normal in our experience now* [italics added]. The profound lesson to me of Dolly was what's normal. What's very normally, ordinarily, within the next few years within the realm of what could be made into a child is changing profoundly. (NBAC, 1999c, p. 108)

Holtzman's observation does not mean that as novel technologies and artifacts become ordinary, they automatically become more ethically acceptable; rather, it means that what constitutes "normal" evolves continually over time. However, in taking a balancing approach to its ethical deliberations (i.e., balancing ethical concerns against the ever-improving medical benefits of scientific advance), and in concluding that the future benefits of embryonic stem cell (ESC) research warrant overriding some people's concerns about treating human embryos in an unethical manner, NBAC does presume that advances in stem cell science and technology will expand the number of morally acceptable ESC sources—and the number of ESC sources appropriate for federal funding. In essence, scientific productivity in ESC research yields the extension of scientific integrity to the use of a wider variety of potentially more controversial ESC sources—the ethical acceptability of stem cell research is coupled to the productivity of that research.

OF EMBRYOS AND OVERSIGHT: THE IMPLICIT MORALITY OF BIOETHICS IN PUBLIC POLICY

NBAC was faced wth two volatile issues in tackling ESC research: the moral status of the embryo, and government funding of controversial research. Over the course of their deliberations, the commissioners chose to take a *public-policy approach* to their analysis, rather than a more incisive ethical analysis. The public-policy approach of bioethics adopts the assumptions of liberalism and abstains from moralism. Thus, the public-policy framing of bioethics asserts that in a pluralistic liberal polity, there is unlikely to be a unified vision about what consititutes "the good life" for a human being, and public policy should not take a position on what the unified vision should be (see chapter 1 of Meilaender, 1995).

NBAC's charge with respect to the stem cell controversy was, from the outset, not well defined and the commissioners were left with some decisions about how to proceed. They were simply asked to "undertake a thorough review of the issues associated with such human stem cell research, balancing all ethical and medical considerations" (NBAC, 1999a, vol. 1, p. 89). However, President Clinton's letter of request also spelled out the need to revisit the thorny issue of human embryo research. Citing federal restrictions on embryo research instituted by his administration, the president reflected that "although the ethical issues have not diminished, it now appears that this research may have real potential" for the treatment of "devastating illnesses" (NBAC, 1999a, vol. 1, p. 89). Beyond this rationale for policy reassessment, the president charge was rather vague, permitting NBAC to consider openly different approaches to its final report, approaches that by commissioners' own admission would have led to different recommendations, as shall be seen. Let us examine now how the commission chose to adopt implicit rather than explicit moral stances on ESC research, first with regard to human embryos, and then to federal funding and oversight.

The Moral Status of the Embryo

In a January 1999 NBAC meeting, chairman Shapiro led a reflection on their charge, clarifying the possible boundaries of their inquiry. On one hand, their "narrowest possible response" was to "look at the issues involved in using existing embryonic stem cells for research purposes," although on the other hand, Shapiro noted, it seemed likely that President Clinton's letter encouraged them to go further: to consider embryo research generally, and even the creation of human embryos for research purposes (NBAC, 1999b, p. 170). However, over the course of the discussion, commissioners came to advocate a narrower, rather than a broader scope. Commissioner Capron, for example, cautioned the other commisioners against trying to cover too much ground in their report, citing the unreasonable workload it would put upon NBAC staff, and agreeing with expert witness Patricia King that too much speculation about future issues was unwise, given the relatively undeveloped scientific knowledge of stem cells (NBAC, 1999b, pp. 173–175). Hence, Capron recommended that they address some version of the question, "Should the Federal regulations relating to research on the embryo and fetus and on pregnant women be amended in light of the new science?" Capron's suggestion addressed the central policy concern in President Clinton's letter, and consequently the struggle between NIH and Congress. NBAC soon elected to concentrate their inquiry on what categories of stem cell research ought to be eligible for federal funding, weighing ethical and medical considerations.

However, in eleventh-hour deliberations, commissioners expressed a curious confusion about the focus of their inquiry, suggesting that they might have felt conflicted about their approach. In particular, there was uncertainty as to whether they were focusing on the ethical issues raised by embryonic stem cell research per se, or those raised by federal funding. In response to commissioner Dumas' assertion that "our major focus is on the ethical issues and implications of the use of stem cells," from which "the issue of federal funding then follows," (NBAC, 1999f, p. 29), Commissioner Greider countered,

> I read this whole report as being very limited to the issue of federal funding…Now if we were to address…the ethical issues irrespective of funding, I would have a very different feeling for the recommendations. I would not come out in the same place that I do. (NBAC, 1999f, p. 32)

Following further debate on the matter, Shapiro asked the commisioners to retrace their footsteps. Citing their educational and policy recommendation roles, he explained that "One cannot deal with...the ethics of federal funding without reminding ourselves what the general ethical issues involved here are" (NBAC, 1999f, p. 37). Shapiro reminded the commission that they had decided early on to focus on the federal funding issue, and not on "what would be morally acceptable for people in the private sector without federal funds to do" (NBAC, 1999f, p. 38). "What we are trying to do here," Shapiro continued, "is recognize that there is moral disagreement out there and trying to design a federal policy that acknowledges the moral worth of other points of view besides our own and reach some kind of compromise"[12] (NBAC, 1999f, p. 40).

NBAC's decision to focus on federal funding of stem cell research clearly marked a shift away from a purely ethical analysis of embryonic stem cell research per se, to an analysis arguably more pertinent to the concerns of the lay public, given the prominence of the abortion debate in late 20th-century American society. Although several commissioners at various times expressed the desire to definitively address the core issue of the human embryo's moral status, particularly during their March 1999 meeting, a *public-policy approach* (eschewing incisive ethical analysis) precluded substantive treatment of the embryo's moral status, further narrowing NBAC's analysis.

Toward the end of NBAC's April 1999 meeting, chairman Harold Shapiro invited the other commissioners to consider the relative merits of discussing a range of views and deriving a middle-ground position, rather than arguing for a particular ethical viewpoint. The commisioners may not agree individually on the morality of the various categories of embryonic stem cell reseach, Shapiro explained, but they could instead recognize "that there are differences of opinion on these issues in our country and we might feel that we have to recommend or should recommend something that is responsive to that fact" (NBAC, 1999e, p. 253). The alternative, Shapiro stated, was "to put forward, for example, a particular moral perspective that we would then have to argue dominates all the others, which, I think, as we all know, would be a difficult task" (NBAC, 1999e, p. 254).

In apparent concordance with the concerns Shapiro raised, the commission avoided making a strong moral argument about the ethical status of embyros. However, NBAC did adopt a moral stance on the issue. Critiquing NBAC for failing to advocate a fair compromise between liberal and conservative viewpoints, John Fletcher (2001) argued that

> When the analogy between permissible abortion and research on HES [human embryonic stem] cells broke down, NBAC turned to urging a benefit:harm ratio. Ultimately it took the position that embryos are forms of human life but not human subjects of research. Whereas the couple who donate gametes are clearly subjects for purposes of research conducted on their embryos, the embryos themselves are not yet fully subjects. In short, NBAC took a stand on the moral status of the embryo, but simply asserted this stand and did not provide convincing argument for it (p. 32).

The commission drew attention away from the fact that it was, in fact, taking a moral stance, by reviewing the different viewpoints on the issue, and justifying its own position as "an intermediate position, one with which many likely would agree" (NBAC, 1999a, vol. 1, p. 50).

The Morality of Federal Funding and Oversight

Having sidestepped incisive moral analysis of the human embryo, NBAC turned its attention to the issue of federal sponsorship of ESC research. As with the moral status of embryos, the commission constructed implicit rather than explicit normative claims. The issue was first raised during a February 1999 discussion about NBAC's agenda, when commissioner Holtzman asked whether the commissioners were in fact going to advocate federal funding. Commissioners Capron and Shapiro replied that no, it was not the commission's job to set research priorities, or to allocate the budget.

However, other commissioners felt that the issue of research priorties was pertient to their inquiry. Commissioner Lo raised the issue again in the April and May 1999 NBAC meetings. In April he asserted that the commission needed better information on the promise of adult stem cell research, in order to assess whether it should get federal funding preference over the more controversial embryonic stem cell research, and to determine the scientific cost of such a preference. Commissioner Miike disagreed, stating that it was not the commision's task to determine such priorities. It was not a matter of placing all bets on one horse, he argued, but rather, the promise of adult stem cell research was pertinent to assessing where NBAC stood on the federal funding of research on stem cells from embryos created expressly for research (NBAC, 1999e, pp. 226–227). Later in that meeting, Shapiro reflected on the commision's focus, stating the need to address federal funding, which was different, for society as a whole, from the general issue. Specifically, he argued, they needed to articulate the arguments for the benefits of federal funding, which was different from arguing that the federal government *should* participate in the research via sponsorship.

In May, Commissioner Lo again raised the issue of research priorities. This time, he acknowledged social justice concerns that had been voiced in a special meeting devoted to religious perspectives, concerns that could be addressed by research funding priorities. Chairman Shapiro responded that their restricted focus on federal funding *eligibility*, rather than advocating particular research priorities, allowed them to put aside distributive justice issues with respect to federal health care research-funding priorities.

Although NBAC concluded that assessing federal research funding priorities and addressing the attendant issues of distributive justice were outside its purview, the commission's argument that certain categories of embryonic stem cell research were *appropriate* for federal funding effectively asserted that those categories of stem cell research *should* be a federal funding priority, because federal funding would enhance the progress of stem cell science, and provide for federal oversight of embryonic stem cell research; the commission made implicit arguments in favor of federal funding, rather than explicit ones.

In order to explain the commission's argument for federal funding, it is necessary to review the functions attributed to federal research funding and oversight over the history of federal bioethics bodies, which have evolved and expanded over the last 3 decades. In reviewing research involving in vitro fertilization and embryo transfer in 1979, the Department of Health, Education, and Welfare (DHEW) Ethics Advisory Board (EAB) noted

> that the procedures may soon be in use in the private sector and that Departmental involvement might help to resolve questions of risk and avoid abuse by encouraging well-designed research by qualified scientists. Such involvement might also help to shape the use of the procedures through regulation and by example. (section D of chapter 6 in EAB, 1979/1988)

Thus, the EAB argued that federal oversight (and by extension, federal funding) promotes integrity and ethical conduct in research, by setting standards and modelling good conduct.

Although the 1988 fetal tissue research panel's report did not express them as explicitly as the DHEW's EAB report did, the considerations of good scientific conduct also figured heavily in the decision of the panel chair, Judge Arlin Adams, to vote for its recommendations in spite of his opposition to abortion. In her NBAC testimony of January 1999, Patricia King recounted Adams's reasoning:

> He says he was able to concur in the report of the HIFFRA [*sic*] panel because he wanted to prevent commercialization of fetal tissue use…He thought that with Federal funding we could employ more careful scientific approaches as well as utilize the highest professional standards, and finally, he thought that without government funding research would be unsupervised and not governed by guidelines. (NBAC, 1999b, p. 110)

In 1999, NBAC gave federal funding and oversight even more moral import in the arguments of its stem cell research report (NBAC, 1999a), expanding the issue to include scientific productivity as well as integrity. In addition to reducing risks and misconduct in research, the conditions attached to public funding "can stipulate that recipients… must share both research results and research materials (including cell lines)" (p. 60) with other researchers. Thus, federal funding "may lead to more widespread dissemination of findings and sharing of materials, which ultimately may enhance scientific discoveries" (p. 60). The final report further noted that "a combination of federal and private sector funding is more likely to produce rapid progress in this field than would private sector funding alone," (p. 59) and "Federal funding is probably required in order for the United States to sustain a leadership position in this increasingly important area of research" (p. 60) (NBAC, 1999, ch. 4).

Making a more emphatic case for the merit of federal funding, commissioner Miike argued, during an NBAC meeting, that

> some form of stem cell research has to be funded by the NIH and Federal Government. And it's not just for the promise of the research, but I think that, and the scientists can correct me if I'm wrong, but I think that this area has such a big promise that if the NIH is not able to fund in this area, they are going to be a defective organization in terms of research as the years go by. They will be shut off from an area of research that is going to be so fundamental to the mission of the NIH, that *it's going to be a defect to their mission* [italics added]. (NBAC, 1999c, p. 80)

From the preceding account, we see that the function of federal oversight and research funding has evolved in the discourse of federal bioethics bodies from that of safeguard and exemplar against misconduct, to acting as a deterrent to commercialization, to being a means to scientific openness and sharing, to the facilitation of rapid scientific progress, to the preservation of the competitive edge of U.S. science, particularly at NIH. Public-policy-based bioethics discourse has legitimated not only regulation, but also public funding, as a means to ensuring ethically sound and productive biomedicine. By endorsing the values of federal funding in such a way, NBAC made implicit normative claims that ESC research should receive federal funding, and although stem cell research was, and still is, in its infancy, the commission assumed that we can count on a future stem cell therapeutic reality.

Furthermore, NBAC's stem cell report, in effect, transforms the public-policy value of scientific productivity and progress into a moral value. When NBAC invited expert witness David Blumenthal to speculate on normative themes that come up in his research, he raised an important but overlooked distinction:

> Well, I sometimes have trouble differentiating between ethical norms and norms that are meant to get the most out of what we're doing, that are more productivity-oriented...People tend to invest academic norms like openness with ethical content. And I don't know whether they are appropriately regarded as ethical norms as opposed to characteristics of universities that render them best suited to furthering public purposes. (NBAC, 1999c, p. 145)

The NBAC commissioners appear to share Blumenthal's difficulty in differentiating between ethical norms and productivity-oriented norms— in fact they conflate them. The technological imperative of ESC research thus becomes not only inevitable, but also ethical.

The previous account has shown that NBAC chose not to pursue incisive philosophical moral analysis about the status of human embryos or government funding of controversial research. At the same time, the commission made implicit normative claims about those issues; it essentially argued that embryos are forms of human life but not human subjects of research, and that the federal government *should* fund ESC research. Through NBAC, bioethics has blurred the boundaries between ethics and public policy, elevated the productivity of science to a moral imperative, and narrowed the debate to minimize issues of general public concern, including the moral status of the embryo, and the just distribution of federally and privately funded biomedical goods in society. These developments raise concerns about the ability of bioethicists and bureaucrats to uphold the integrity of science when it comes into conflict with the productivity of science, and to define integrity and productivity in terms of the public good as well as the interests of the scientific enterprise.

STANDARDIZED PACKAGES: NBAC'S RECOMMENDATIONS

Neither the science of stem cell research nor the consequent therapeutic technology is certain, and thus, neither is the morality of unfettered

ESC research. Further ethical uncertainty is generated by the particular circumstances of each ESC derivation, and the particulars of each research study. In such situations, respectful, ethical treatment of embryos, ESCs, and human research subjects requires regulatory assurance. To deal with this inherent uncertainty, NBAC customized several existing standardized packages from the public bioethics toolbox for its recommendations, which were presented in its September 1999 report, *Ethical Issues in Human Stem Cell Research* (NBAC, 1999a).

Firstly, NBAC began by articulating that certain types of ESC research should be *eligible* for federal funding, based on the sources of ESCs used in the research; cadaveric tissue and excess embryos from infertility treatment were deemed to be ethical sources of ESCs for research. Federal funding would enable the imposition of federal regulation to assure that research met ethical standards. The rules and procedures required by the federal regulations, and the funding mechanisms that require their use, constitute standardized packages, which serve to stabilize tensions between science and politics/ethics. Over the last 3 decades, federal bioethics bodies have promoted variously federal funding of controversial embryo research, as a safeguard and exemplar against scientific misconduct, a deterrent to inappropriate commodification, and a means to scientific openness, rapid scientific progress, and U.S. technological (and thus economic) competitiveness (see Leinhos, 2002b). Indeed, NBAC cited nearly all of these as arguments in favor of selected federal funding of ESC research.

Secondly, NBAC recommended detailed informed consent and collection practices for the donation of so-called excess IVF embryos, modeled on organ and fetal tissue donation guidelines, in order to guard against coercion and other inappropriate incentives to donate.

Thirdly, NBAC carefully outlined a three-tiered oversight system for federally funded ESC research, in the interest of dividing labor while avoiding the moral hazards of self-policing. At the topmost level, a National Stem Cell Oversight and Review Panel (National Panel), housed at the Department of Health and Human Services (DHHS), was to develop standard guidelines for protocol review at all levels, and to coordinate a variety of integrity assurance mechanisms (which shall be discussed later). At the local level, institutional review boards would assure researchers' compliance with the National Panel's guidelines and other existing ethical guidelines for research conduct. To guarantee uniform ethical practice at government agencies, occupying the middle tier, those agencies sponsoring ESC research would ensure that agency-level review complied with the National Panel's requirements. Through the peer review process, agencies would give special attention to adequate justification for ESC use, as signified by each project's scientific merit.

Fourth and finally, NBAC called for voluntary compliance with and participation in the federal oversight mechanisms by privately funded researchers conducting research eligible for public funding, and for voluntary development of and compliance with private industry research safeguards and standards for research ineligible for public funding (e.g., research with ESCs derived through SCNT techniques). Taken together, these four standardized packages provide a system of bioethical tools for managing the ethical uncertainty of ESC research while allowing it to progress with close monitoring.

NBAC's stem cell research recommendations, not surprisingly, consistently serve the mutual interests of science and policy makers, even though those policy makers occupy different branches of government, and represent interests that at times compete. An examination of the proposed responsibilities of the National Panel illustrates how the panel serves the interests of both science and policy makers. Firstly, the National Panel was supposed to review ESC derivation protocols, approving acceptable ones and certifying cell lines derived through them. Secondly, the National Panel was supposed to maintain a national public registry of approved protocols and certified cell lines, and to link the registry to a panel-sponsored public database of identifying information, data, and publications connected to approved protocols and cell lines. Finally, the panel was supposed to use the registry and database to track the history and use of certified cell lines, and to report periodically to the DHHS secretary on the state of stem cell science and its attendant ethical concerns, reevaluating the continued adequacy of NBAC's regulatory recommendations.

Clearly, the activities proposed for the National Panel would not only provide means to and indicators of the assurance of scientific integrity in ESC research, but also would coordinate and facilitate the entirety of the American stem cell research enterprise by providing a centralized database of information on stem cell research. That is, the National Panel would function to assure both the integrity and productivity of stem cell science, things mutually desired by the scientific community and the federal government. The National Panel's proposed activities would also serve to legitimize the field of bioethics as a mediator, by continuing to call for the services of bioethics experts, and by perpetuating the standardized packages developed in collaboration with those experts.

The measures that NBAC recommended were crafted to manage the scientific and ethical uncertainties that contribute to the tension of the stem cell research controversy. If the recommendations had been fully implemented, the impact of uncertainty would have been controlled and diminished through the use of standardized packages, by breaking the uncertainty into bite-sized procedures, protocol by protocol.

However, NBAC lacked any effective standardized packages for dealing with the uncertainty of how their recommendations would be received, and whether, in fact, they would be implemented. Many government organizations, particularly those with relatively short track records, face similar uncertainties and are susceptible to the direction in which the political winds are blowing. Advisory bodies, however, appear to be particularly insecure in their ability to succeed as boundary organizations. Their ability to stabilize and contain the science-politics boundary at the site of controversy is often limited to the time frame during which the bodies' deliberations occur, unless their recommendations are implemented, or they are given some "action-enforcing power"—the authority to command either the acceptance of their recommendations by an executive official, or to require a public accounting of reasons for the official's rejection of the recommendations (Bulger et al., 1995).

Although the Clinton Administration immediately distanced itself from NBAC's recommendations (Marshall, 1999), it appears that NBAC may have had some influence on the decision for NIH to implement a stem cell registry similar to that recommended by NBAC. Like the Human Embryo Research Panel before it, NBAC incited prolife sensibilities by recommending a relaxation of federal embryo research restrictions, and contributed to congressional foment on the issue. Ironically, this may have had the effect of strengthening NIH's position in the political struggle, by making its advocacy of limited embryonic stem cell research appear relatively moderate (Fallows, 1999). In the end, NBAC may have strengthened arguments in support of stem cell research, making it politically impossible for President Bush to prohibit stem cell research altogether 2 years later when he announced that publicly funded research would be permitted on a limited number of ESC lines specified by his administration. In a less conservative political climate (or perhaps at some future date, as implied by commissioner Holtzman), NBAC's recommendations might have been implemented, and subsequently might have stabilized the science-politics boundary effectively through the execution of the various standardized packages advocated.

CONCLUSION

This chapter has examined boundary work performed by NBAC in the controversy surrounding embryonic stem cell research, at the boundaries between science and ethics, and between ethics and public policy. It described the coupling of scientific and ethical uncertainty, and the coupling of research productivity and integrity assurance at these borders in the commission's deliberations on embryonic stem cell research, and it argued that NBAC's boundary work ultimately served to reinforce the authority of science and to marginalize the contrary moral concerns of some citizens.

The analysis in this chapter indicates that the adoption of a public-policy approach in the bioethical argumentation of advisory bodies has significant implications for the nature of the arguments and recommendations produced by those advisory bodies. If policy makers frame the charges of advisory bodies for the purposes of political conflict management, rather than purposes of genuine ethical inquiry, the output of advisory bodies may well limit debate in ways that privilege some societal interests over others. The public good would perhaps be better served by thoughtful construction of advisory-body charges that specify inquiry of difficult questions, and attention to a variety of perspectives on those questions.

From an STS perspective, the chilly reception of NBAC's stem cell research recommendations might be viewed as something of a default victory for participatory science policy, rather than as being a boundary organization failure. Ideally, participatory science policy fosters access and inclusions of civic perspectives in the policy process, beyond the institutionalized elite viewpoints that dominate science policy, resulting in more balanced and informed public debate and decision making (Kelly, 2003). Policy decisions subsequent to the release of NBAC's recommendations, that take a more cautious and conservative approach to ESC research, reflect the moral reservations of many citizens. President Bush's decision to allow some federally funded research on a limited number of established cell lines managed to offend both moral conservatives and the scientific community (since the available cell lines could not provide the basis for robust science).

Federal bioethics bodies might more effectively serve the public good and the multiple interests it represents by resisting political pressure and pursuing richer ethical inquiries. In a step toward such an approach, the charter of current President's Council on Bioethics (the Council) stipulates that "the Council shall be guided by the need to articulate fully the complex and often competing moral positions on any given issue, rather than by an overriding concern to find consensus" (Executive Order 13237, 2001, sec. 2c). However, the appointed roster of the Council includes only academics and no laypersons, placing the onus on the Council itself to seek and incorporate broad-based public input into its proceedings. Furthermore, accusations by a former Council member that the science presented in the Council's report on stem cell research was tainted by political bias suggest that the current iteration of public bioethics in the United States is not succeeding in meeting the ideal of participatory science policy (Blackburn, 2004; see also related letters to the editor in July 15, 2004, issue of the *New England Journal of Medicine*).

Guston (2000) suggested that boundary work performed outside the confines of a boundary organization may be "worrisome," because, "there is nothing to prevent the boundary work, necessarily laden with interests, from being self-serving to the extreme. Outside of boundary organizations, outcomes are therefore more likely to be determined by the exercise of power" (p. 152). Because boundary organizations are beholden to two sets of principals (in the case of NBAC, scientists, on the one hand, and policy makers, on the other) with overlapping but different aspirations, they can be expected to stabilize the relationship between science and politics without favoring one over the other. However, to the extent that science and politics exercise joint power to achieve mutual goals, the boundary work performed within science-politics boundary organizations may also be determined by the exercise of power, to the exclusion of interests outside the state and the scientific community. Furthermore, boundary organizations may serve to protect that exercise of power, assisted by the specialized mediators who have constructed a professional jurisdiction by serving the mutual interests of their political and scientific principals.

By arguing, in effect, that the biomedical benefit of stem cell science outweighs concerns about strong respect for embryos, NBAC coupled the concepts of scientific integrity and productivity as they apply to stem cell research, and combined the assurance functions for both. The risk of this combination is that productivity concerns may come to overshadow other values that are associated with scientific integrity in American society. The continued reliance of public bioethics on expert committees, which privilege the scientific model of knowledge production and argumentation, is unlikely to yield science policy that captures the full richness of societal debate on bioethical controversies.

ENDNOTES

1. Acknowledging scholarly debate about the nature and significance of the concept, Jonsen (1998a) defined ethos as:

    the characteristic way in which a people interpret their history, their social world, and their physical environment in order to formulate convictions and opinions about what is good and right. Ethos is not the actual behavior of a people... . It is the panorama of ideas and ideals by which a people judge themselves... . Nor is ethos the collection of rules and principles that are invoked. It is the matrix in which those rules principles and values are formed. (p. 389)

2. Here, I rely heavily on Bledsoe's (1997) overview of the nature and operation of presidential commissions.

3. The ill-fated Biomedical Ethics Advisory Committee (BEAC) was created by Congress in 1985. It took a year for Congress to select a board of congresspersons whose job it was to appoint the BEAC members; subsequently abortion politics deadlocked the board and quashed the advisory committee. BEAC operated from 1988 to 1989, meeting only twice and issuing no reports.

4. Commissioner Charo recused herself as of February 1999, to avoid the appearance of conflict of interest due to her role at the University of Wisconsin in making some recommendations about the work of stem cell researcher Jamie Thompson at the University of Wisconsin. Commissioner Greider recused herself in July 1999, when the stem cell report and recommendations were virtually complete; here again, the intent was to avoid the appearance of conflict of interest due to her employment at Johns Hopkins University, which has a direct financial interest in the stem cell research of John Gearhardt.

5. These critics do not come from a world that NBAC is primarily answerable to; as we shall see, critics from religious and other pro-life organizations are marginalized in NBAC's stem cell deliberations.

6. With regard to political affiliation, I was unable to identify any Republicans among the commissioners, but I did identify three Democrats.

7. Scientific issues and testimony occur first in both the deliberations and in the final report, whereas religious testimony and viewpoints were among the last heard, are primarily relegated to a summary in an appendix of

volume 1, and are presented at length only in the third and final report volume.

8. The properties of these cells will be discussed in the next section of this book.

9. The cells cultured by Thomson's and Gearhart's research groups are designated respectively as embryonic stem cells and embryonic germ cells. For ease of reference, I refer to them here collectively as embryonic stem cells, or ESCs.

10. At the time of NBAC's deliberations, there was disagreement among developmental biologists (and in other circles) as to the precise meanings of the terms *totipotent* and *pluripotent*. Some defined totipotency as the ability to develop into a complete organism, and pluripotency as the ability to develop into any of the various cell types found in an adult, but without the ability to develop into an entire organism. Others defined totipotency as the ability to differentiate into any cell type (but not necessarily the entire organism), and pluripotency as the ability to generate more than two different cell types. More recently, the term multipotency has emerged in stem cell discourse. Development and debate of this terminology is beyond the scope of the present essay; for present purposes the reader may simply think of totipotent cells as having more developmental options than pluripotent cells.

11. Applying the cloning technique (SCNT) to stem cell research is one approach being pursued, with the objective of creating Molly, the cloned stem cell, rather than Dolly, the cloned ewe, by inserting a differentiated cell nucleus into an enucleated stem cell. There has been an attempt to distinguish this use of cloning as *therapeutic* cloning, but this distinction has not yet succeeded in allaying fears that researchers might attempt to implant an SCNT product into a woman's uterus and produce a baby.

12. Following his lengthy clarification, Shapiro asked the commission whether they still agreed to the funding approach; several commissioners conveyed their agreement, and none dissented. My account in this essay should not be taken as an indication that Shapiro pressured the commission into taking a particular approach; Shapiro largely restricted himself to the role of facilitator, here paraphrasing and summarizing the commissioners' discussion for the purpose of clarifying and advancing their deliberations.

CHAPTER 7

CONCLUSION

The analysis contained in the preceding chapters makes it clear that the question whether bioethics is a lapdog or a watchdog to the biomedical enterprise, in addition to being a false dilemma, fails to apprehend the myriad social intricacies that constrain the capacity of bioethics to develop and promulgate a rigorous critique of biomedical research and practice. Importantly, the sociostructural framework within which bioethics has emerged and grown includes the funding and policy climate that institutions of higher education must weather; the professional culture and norms of academic biomedicine into which academic bioethicists are expected to assimilate; the particular expertise for which bioethicists might be held legally accountable in the courtroom; and the particular philosophical and political logic through which bioethics must offer public-policy advice.

Part I observed that academic bioethics is viewed by university administrators as a means to tap into opportunity structures and to protect or legitimate biomedical research, at least in the case of Letters University. However, ethics education requirements for medical education accreditation on one hand, and for grant funding from federal agencies on the other hand, also compelled Letters University to beef up bioethics, and medical school faculty and students appeared to be genuinely appreciative of having access to bioethics. On a deeper level, it also recognized that isomorphic pressures can have a pervasive influence on the way that academic bioethics "does business"—by virtue of being expected to perform to the same revenue-generating and promotion and tenure standards as other units and faculty in the medical school, the Bioethics Center at Letters University and its faculty members are substantially shoehorned into the knowledge-production model of the academic medical center they call home. Furthermore, close association with academic medical culture and structure likely makes it more difficult for faculty members at the Center to forge effective relations with constituents such as the nursing profession and marginalized extramural communities.

The strained relations between the Center and the IRBs at Letters University attest to the motivation of Center faculty to assure the ethical integrity of human subjects research and the protection of participants. As research litigation increases, so does the opportunity for academic bioethicists to contribute to research ethics, both in IRB meetings and in the courtroom. However, in order to do so, they must determine how to interact effectively with the actors and structure of the IRB oversight system, which are as firmly rooted in the research culture and infrastructure of the academy as they are in the federal regulations mandating their existence.

In the realm of law, Part II observed that the bioethical expertise considered relevant to biomedical research and practice is governed by federal and state regulations, tort theories, the case law of medical malpractice, the nature of the advice sought by medical practitioners and researchers, and the perceptions of insurance companies that might indemnify ethics consultants. Health care ethics consultants appear to have considerable opportunity for input into what constitutes their accountable expertise by means of defining and promulgating formal core competencies and a code of ethics, which would be legitimated by the courts as a documented peer standard of care in any tort cases heard against ethics consultants. However, the reluctance of bioethicists to demarcate firmly their expertise, due to the highly prized interdisciplinarity of the field, a liberalism-based aversion to claiming moral authority, and goals more aligned with an ideology of social trustee professionalism than an ideology of expert professionalism, make it unlikely that bioethics experts will capitalize on this potential.

The strength of the professional jurisdiction of bioethics is arguably based more on external legitimacy *constructed by other actors*, whether opportunistically or morally founded, than it is on conscious strategic attempts of bioethicists to control or build their jurisdictional claim. As a relatively weak advisory jurisdiction to medicine, science, and increasingly, law, the usefulness and therefore much of the legitimacy of bioethical expertise is constructed by the needs and motives of these elite professional groups.

Graduate education in bioethics is a major source of external legitimacy for the field, particularly for establishing academic credibility within the academy. MBE programs provide formal legitimacy by signifying that bioethics is a "real" academic field, and by generating tuition revenues for resource-limited postsecondary institutions. They also reinforce a cohesive and organized identity for the field, through the institutional isomorphism that shapes and homogenizes these MBE programs, and by providing degree recipients with concrete, explicit evidence that they are, indeed, legitimate bioethicists.

In the public-policy arena of executive-branch bioethics advisory bodies, the public discourse of bioethics is constrained by the dual Weberian rationalization imposed by the institution of science, on the one hand, and by the liberalism-based public-policy approach of the state, on the other. The outlook for an autonomous and influential policy-advising voice for bioethics is made all the more bleak by the close-knit interdependence of science and the state in modern U.S. society, fueled by federal science policy and culminating largely in the knowledge-production engine of academia, which is also the primary institutional headquarters of bioethics. The presence of bioethics in federal public policy has been legitimated by the institutions of executive advisory bodies and oversight procedures, but at the cost of incisive moral analysis of the truly divisive issues of the day.

IMPLICATIONS FOR THEORY OF PROFESSIONS

Abbott (1988) argued that the primary limitation of the theoretical concept of professionalization is its focus on structure rather than professional work, and he employed the system of professions as the central unit of analysis. In keeping with this, I have conceptualized the work of professions primarily as boundary work, and examined the boundary work performed by bioethicists in the workplace, legal, and public arenas. Additionally, I have relied upon organizational theory to develop an account of the distinctive institutional constraints governing the boundary work of bioethics, which must engage with the institutions of academe and the state, institutions that produce and regulate the biomedical research and practice which bioethics purports to advise.

Abbott (1988, 1991) has also chastised the tendency to reify professions and professionalization, which obfuscates some of the intricacies of expertise, power, and work in the social world. The work presented here provides more reason not to essentialize professions. Bioethicists, although seeking the power to influence biomedical research and practice, are explicitly ambivalent about professionalization, as is evidenced by their discourse, and by what is *not* in their discourse, namely, required competencies, accreditation, certification, and a professional code of ethics. Bioethics is intentionally and proudly interdisciplinary, nonexclusionary, and antiauthoritarian, though in an inverted sense; bioethicists are against their *own* (moral) authority, more than they are against the (expert) authority of others.

Furthermore, while bioethics does, at some level, seek to limit the authority of biomedical science when it violates ethical norms, its "biotechnofetishism" belies the underlying assumption that bioscience can, and perhaps should, leave no stone unturned on the road of progress. Perhaps more unfortunately, bioethicists' general penchant for genetic gadgetry overshadows more mundane, and far more pervasive ethical challenges in medical care and research (such as achieving a more just distribution of health care resources nationally and globally) that would arguably benefit from more bioethical scrutiny.

The preceding chapters also reminded us that there is more to look for in the system of professions than jurisdictional contests. One searches in vain to find any other professional groups competing for the jurisdiction of bioethics, although the clergy would perhaps compete if they had more authority in the contemporary era. The lack of challengers to bioethics is partly due to the fact that all newcomers to the jurisdiction are welcomed into the bioethics community with open arms; the only qualifications are deep firsthand familiarity with biomedical practice, and one or more of several expertise sets that illuminate the social challenges that arise in biomedicine.

With these factors established, prospective competitors are merely assimilated into the bioethics melting pot. Philosophers and social scientists, to the extent that they are able to overcome their own aversion to mucking about in real-world problems and making normative claims, may venture into the bioethics discourse from the comfort of their disciplinary homes, at the risk of derision from their departmental colleagues. However, more important than the lack of serious contenders for the jurisdiction of bioethics is a distinctly symbiotic relationship between the bioethics and biomedical jurisdictions. The institutions of bioscience and medicine, as well as the state, are well served by sustaining the advisory jurisdiction of bioethics, because it serves to buffer them from politics and controversy, and to legitimate their activities. In this respect, the bioethics jurisdiction serves as a sort of boundary organization as described by Moore (1996), one that enables biomedicine to simultaneously maintain credibility as a scientific enterprise and as a professional community serving the public good in an ethical manner.

Abbott's account of professionalization does not adequately characterize the state of professionalization or discipline formation that bioethics represents and shares with the field of higher education studies (Abbott, 1988). Neither field has become a full-fledged profession, nor developed a distinct disciplinary identity, but rather they have persisted in a quasi-professional state with no culminating metamorphosis in sight. Perhaps they instead represent a new class of emerging interdisciplines that develop flexible expertise to train the flexible workforce of the new economy. These fields, developing during the ascendance of the academic capitalism knowledge/learning regime, draw from various knowledge sets to forge hybrid expertise in order to address problems stemming from the simplification of knowledge into a raw material for economic productivity. However, bioethics is still rooted in the public-good knowledge/learning regime, but grappling with the purposes of both the academic-capitalism and public-good knowledge/learning regimes.

In Abbotts' theoretical framework (1988), professional legitimacy in the legal arena is quintessentially associated with successful lobbying for state licensure of a profession, resulting in the privilege of substantial self-regulation. However, there are a few other ways in which professional groups can achieve legitimacy in the legal arena that are deserving of further study. In the contemporary litigious society, professional liability ironically presents itself as an form of legitimacy worthy of exploration. Claims of professional negligence or malpractice must be proven in court by demonstrating that a professional standard of care has been breached. The professional standard of care is a matter of expert knowledge, evidenced by formal professional discourse and the testimony of expert witnesses. In addition to liability, legal legitimacy of expertise may be conferred by several means, including statutes which encourage or require the use of expertise in certain circumstances, such as the required use of ethics consultants in controversial patient-care decisions; the codification of certain expert concepts into regulations, such as the requirements of informed consent in the Common Rule; and statutes conferring special immunity to certain expert decision makers in service to government institutions. Examining boundary work aimed at establishing expertise related to these claims in the legal arena can provide insights into the legitimacy and logic of the expertise in question.

Bioethics and the Fourth Estate: Bane and Boon

Noting the greater per capita quantity of radios and televisions in the United States, as compared to Britain and France, Abbott (1988) construed that "pervasive media have thus kept the public arena central for professional claims in the United States" (p. 165). The role of the media cannot be overlooked in an account of the logic and legitimacy of bioethics in the United States.[1] In Part II, we heard concerns expressed by established and future bioethicists that the field is too media driven, and it is important to reflect on the way that media presence both legitimates and shapes bioethics.

The media represents a double-edged sword for bioethics. On the one hand, attending to current events does establish the relevance of bioethics to the public, and more importantly, provides a powerful megaphone for the voice of bioethics (Simonson, 2002; Hopkins, 1998). This megaphone can be wielded defensively as well as offensively, affording bioethicists a limited amount of protection from reprisal, on pain of bad press for the retaliator. On the other hand, a media orientation has a negative impact on academic credibility, as well as the reflexive capacity of the field, due to the reactionary, sensationalist mode of knowledge production in which it results. Furthermore, ethicists who engage with the media are subject to the constraints of journalism, the most significant of these being the need avoid offending sponsors and corporate interlocks.[2]

Acknowledging that journalism constitutes "a special and crucial sub-region of the policy process," Goodman (1999, p. 182) criticized journalists' common use of ethicists to pass moral judgment on news issues, and argued that journalists and ethicists have a responsibility to cover issues and arguments collaboratively and without oversimplification, and to present intelligent disagreements without indulging in moralizing. Nonetheless, even poor media coverage has value to bioethics; Simonson (2002) contended, "bioethics can benefit from even the most sensationalized, sound-bitten and superficial portrayal. Every time such a story appears, the public significance of bioethical issues is reconfirmed" (p. 35). The boundary between bioethics and journalism is clearly worthy of attention by sociologists of bioethics, and of reflection by bioethicists.

IMPLICATIONS FOR HIGHER EDUCATION STUDIES (HES) AND SCIENCE AND TECHNOLOGY STUDIES (STS)

The research described in this book contributes to HES and STS in several respects. Firstly, I have focused attention on several aspects of knowledge production, a subject rarely addressed in HES. By conducting a case study of the development of an academic bioethics C&I, I have paid attention to discipline formation and examined the engagement of faculty with institutional pressures that shape their work, illustrating how the academic-capitalism knowledge/learning regime impacts the knowledge faculty members produce.

This case study also sheds some light on the ways in which knowledge produced in the academy is linked to the larger culture in the legal and public arenas. By demonstrating how bioethics is affected by academic capitalism, and how it also engages in academic capitalism, I extend Slaughter and Rhoades' (2004) analysis of academic capitalism beyond the traditional physical science and high-technology fields that are linked more directly to the new economy. Bioethics may be viewed as an inter-stitially emergent function, cultivated not to manage activities directly related to the generation of external revenues, but rather to manage the political and social implications of revenue-generating biomedical sci-ence activities. Rather than technology transfer, bioethics manages the cultural transfer of scientific results.

With regard to STS, my analysis places emphasis on the institutional conditions of knowledge production, and postsecondary educational institutions in particular, which have received little attention in STS. I contend that bioethics is worthy of serious scholarly attention, and is not "just politics," as some STS scholars have suggested (Leinhos, 2002a). While it is true that bioethics is political, simply saying so, without examining the ways in which bioethics is political and why it is so, fails to appreciate the complex challenge of seeking institutional legitimacy while simultaneously working to speak truth to power. Bioethics faces significant institutional pressures as it seeks to advance social welfare goals in the new economy, and, as its products, such as the informed-consent process, are co-opted by stakeholders outside of bioethics for various purposes (e.g., institutional use of consent forms as a protection against legal liability).

In fact, STS shares several instructive similarities with bioethics. Both have established interdisciplinary, weak, advisory jurisdictions to science, and both were initially accepted in this role by the scientific community on the premise that these fields might serve as boosters to the institution of science, and defuse some political criticisms of science (Winner, 2001). The "science wars" have soured the relations between STS and the science community, possibly permanently. Will bioethics eventually suffer from the same rift? At this time that seems unlikely, given that bioethics is more regularly accused of being too intimate with biomedical corporations than of being too critical of biomedicine. This possible eventuality is made even less likely by the close relationship of bioethics to medical education and practice. Bioethicists must immerse themselves in the culture and assumptions of medicine in order to provide meaningful and valuable expertise to medical professionals, and they face disincentives to criticize that culture in a systematic way.

THE LEGITIMACY AND LOGIC OF BIOETHICS

In order to influence even a small aspect of the social world, one has to participate in it. Achieving influence requires legitimacy, which can only be obtained by playing a part in existing institutions, cultures, and power structures, something which, in turn, means being shaped by those institutions and power structures, throughout the social nexus.

The predicament of bioethics has some strong parallels to the challenge faced by AIDS activists in impacting AIDS research and drug development. In order to achieve credibility with the research establishment, activists needed to become scientific experts. However, developing scientific expertise exposes activists to the lure of science and the seduction of knowledge and power, and inevitably moves them closer to the scientific worldview. "Ironically," Epstein (1996) observed, "insofar as activists start thinking like scientists and not like patients, the ground for their unique contributions to the science of clinical trials may be in jeopardy of erosion" (p. 342). Some thoughtful activists are "reflexively engaged" (p. 342) in addressing such issues, and acknowledge the value of continual connection with the more visceral perspective of persons living with AIDS (Epstein, 1996).

Compared with these AIDS activists, bioethicists face a considerably more difficult challenge establishing some reflexive distance from the seduction of biomedicine. Whereas many AIDS activists are themselves persons living with AIDS, and are thus dependent on AIDS research that both develops effective treatments and accounts for the human needs of AIDS patients, the dependency of bioethics on biomedicine takes a less explicit, less visceral form. The legitimacy of bioethics depends on its ability to produce something of value to science, medicine, and the state, including both useful, substantive advice, and the ethical authorization of scientific and medical progress. If bioethics is to earn legitimacy from the public as a watchdog, it must leverage the power of the media judiciously, resist the reactionary, sensationalist mode of knowledge production, maintain some reflexive distance from biomedical professions, and build close ties with its original constituents: patients, research participants, and marginalized communities. For those who inhabit the headquarters of the elite ivory tower, this is a tall order indeed.

ENDNOTES

1. In the information age, the relationship of bioethics to the media arguably deserves attention in other nations as well. For an account of bioethics in the German media regarding preimplantation genetic diagnosis, see Graumann (2000).
2. For an account of the constraints faced by science journalists, see Nelkin (1995).

APPENDIX

METHODOLOGY:
NBAC's DELIBERATIONS
ON EMBRYONIC STEM CELL RESEARCH

Over the past 3 decades, the U.S. federal government has called upon bioethics experts to assist in the navigation of policy dilemmas posed by new biomedical technologies and their societal context. Several bioethics commissions have provided bioethics with a formal role in U.S. federal governance, the most recent being the National Bioethics Advisory Commission (NBAC), created by President Clinton in 1995. The NBAC's charter expired in October 2001; in August 2001 President Bush established a council initially devoted entirely to human stem cell research, to be headed by bioethicist Leon Kass.

Compared with the judicial forum, federal advisory commissions provide a more prominent role for bioethicists in constructing boundaries of their own expertise and the expertise of others. However, the roles of bioethicists are here constrained by their organizational structure: a presidential advisory commission. How do NBAC commissioners socially construct themselves and other constituencies in the policy arena? How does boundary work performed by NBAC commissioners and their discourse affect the legitimacy and logic of bioethics? In what ways does the NBAC itself constitute a boundary organization? To what extent do the ideals of the original bioethics social movement remain intact?

This study used the published reports, meeting transcripts, funding data, and charter of the NBAC (much of which is available online at www.bioethics.gov), as well as commissioners' outside publications, secondary literature about the NBAC and prior bioethics commissions, and the more general literature on presidential commissions. Biographical career data of commissioners was also collected.

I conducted discourse analysis on the NBAC meeting transcripts and reports of relating to the stem cell controversy. To guide my analysis of the meeting transcripts, I generated summary forms and read the transcripts carefully, using categories and questions derived from my research questions. I also took detailed notes of the meeting proceedings and analyzed them for emergent themes and boundary work.

Several features of the NBAC make it a suitable research site for studying federal-policy discourse. Bioethics issues are the focus of various federal activities, particularly at the National Institutes of Health. However, NBAC is a high-profile setting, issuing reports and recommendations on several issues. The historical context of NBAC includes other centrally located federal bioethics commissions (including the influential President's Commission), each with different effects on public policy, providing a comparative context for analysis of NBAC activities. NBAC commissioners come from various social spheres, and include several high-profile academic bioethicists, providing an analytic link between academic and government activities and concerns. While the advisory-only status of the NBAC is potentially detrimental to its policy impact, the nonbinding nature of its recommendations provides an opportunity to examine the treatment of bioethics issues in the executive branch, treatment which is visible in the extent to which recommendations translate into action or codified procedures. Nonbinding recommendations also maximize opportunities for other stakeholders to co-opt or critique recommendations and justifications according to their own purposes, which makes such recommendations particularly useful for the study of legitimacy and knowledge production.

This investigation focused on the issue of human stem cell research in NBAC discourse. Like the issue of human cloning, stem cell research is rich in both scientific and ethical detail. However, stem cell research has more extensive medical applications than human cloning does, and its realization as a useable biotechnology is closer to fruition than human cloning's is. Thus stem cell research has promising commercial implications, and the policy discourse can be examined for the interactions of commercial pressure with bioethics legitimacy and concepts. In this way, stem cell research provides a model controversy for examining boundary work performed by scientists and ethicists, as they negotiate the tensions of academic and entrepreneurial interests.

REFERENCES

AAHRPP (Association for the Accreditation of Human Research Protection Programs, Inc.). (2008, March 20). *AAHRPP accredits 15 research organizations, bringing total to 107*. Press release. Retrieved April 19, 2008, from http://aahrpp.org/www.aspx?PageID=234.

Abbott, A. (1988). *The system of professions: An essay on the division of expert labor*. Chicago: The University of Chicago Press.

Abbott, A. (1991). The order of professionalization. *Work and Occupations, 18*(4), 355–84.

Allen, A. (1996). The march through institutions: Women's studies in the United States and West and East Germany. *Signs, 22*(1), 152–80.

Anspach, R. (1993). *Deciding who lives: Fateful choices in the intensive-care nursery*. Berkeley, CA: University of California Press.

APA (American Philosophical Association). (1997). 1997 survey of programs in bioethics. *APA Newsletter on Philosophy and Medicine, 96*(2), 80–102.

ASBH (American Society for Bioethics and Humanities). (1998). *Core competencies for health care ethics consultation*. Report of the ASBH Task Force on Standards for Bioethics Consultation. Glenview, IL: Author.

ASBH. (2001). *North American graduate bioethics and medical humanities training program survey*. Conducted by the ASBH Status of the Field Committee. Glenview, IL: Author.

ASBH. (2004). *ASBH task force report on ethics consultation liability*. Retrieved October 24, 2005, from http://www.asbh.org/resources/taskforce/index.htm.

ASBH. (2007). *Fall newsletter*. Retrieved January 21, 2008, from www.asbh.org/publications/pdfs/fallseven.pdf.

Ashmore, M., Edwards, G., & Potter, J. (1994). The bottom line: The rhetoric of reality demonstrations. *Configurations, 1*, 1–14.

Aulisio, M., & Rothenberg, L. (2002). Bioethics, medical humanities, and the future of the 'field': Reflections on the results of the ASBH survey of North American graduate bioethics/medical humanities training programs. *American Journal of Bioethics, 2*(4), 3–9.

Baker, R. (2005). A draft model aggregated code of ethics for bioethicists. *American Journal of Bioethics, 5*(5), 33–41.

Bardon, A. (2004). Ethics education and value prioritization among members of U.S. hospital ethics committees. *Kennedy Institute of Ethics Journal, 14*(4), 395–406.

Beauchamp, T., & Childress, J. (1994). *Principles of biomedical ethics*. New York: Oxford University Press.

Bell, D. (1969). Government by commission. In T. Cronin & S. Greenberg (Eds.), *The presidential advisory system* (pp. 117–123). New York: Harper and Row.

Beller, G. (2000). President's page: Academic health centers: The making of a crisis and potential remedies. *Journal of the American College of Cardiology, 36*(4), 1428–1431.

Bevins, M. (2002). Why medical humanities? *American Journal of Bioethics, 2*(4). Retrieved October 31, 2005, from http://taylorandfrancis. metapress.com/link.asp?id=6bu3keq9vunfavxw

Blackburn, E. (2004). Bioethics and the political distortion of biomedical science. *New England Journal of Medicine, 350*, 1379–1380.

Bledsoe, W. C. (1997). Presidential commissions. In *Cabinets and counselors: The President and the executive branch* (2nd ed., pp. 181–195). Washington, DC: Congressional Quarterly Inc.

Bloor, D. (1994). A sociological theory of objectivity. In S. Brown (Ed.), *Objectivity and cultural divergence* (pp. 229–245). New York: Cambridge University Press.

Bosk, C. (1979). *Forgive and remember: Managing medical failure.* Chicago: University of Chicago Press.

Bosk, C. (1992). *All God's mistakes: Genetic counseling in a pediatric hospital.* Chicago: University of Chicago Press.

Bosk, C. (1999). Professional ethicist available: Logical, secular, friendly. *Journal of the American Academy of Arts and Sciences, 128*(4), 47–68.

Bosk, C. (2002). Now that we have the data, what was the question? *American Journal of Bioethics, 2*(4), 21–23.

Bosk, C. (2003). The licensing and certification of ethics consultants: What part of "No!" was so hard to understand? In M. Aulisio, R. Arnold, & S. Younger (Eds.), *Ethics consultation: From theory to practice* (pp. 147–164). Baltimore: Johns Hopkins University Press.

Bosk, C., & Frader, J. (1998). Institutional ethics committees: Sociological oxymoron, empirical black box. In R. DeVries & J. Subedi (Eds.), *Bioethics and society: Constructing the ethical enterprise* (pp. 94–116). Upper Saddle River, NJ: Prentice Hall.

Bouvia v. Superior Court, 225 Cal. Rptr. 297, 297 (Cal. Ct. App. 1986).

Boyce, N. (2002). A view from the fourth estate. *Hastings Center Report, 32*(3), 16–17.

Boyce, N., & Kaplan, D. (2001). And now, ethics for sale? *U.S. News & World Report, 131*(4), 18.

Brainard, J. (2004, October 22). Lobbying to bring home the bacon. *Chronicle of Higher Education*, p. A26.

Brennan, T. (1992). Quality of clinical ethics consultation. *Quality Review Bulletin, 18*(1), 4–5.

Brint, S. (1994). *In an age of experts: The changing role of professionals in politics and public life.* Princeton, NJ: Princeton University Press.

Brock, D. (1987, July). Truth or consequences: The role of philosophers in policy-making. *Ethics, 97*, 786–791.

Brody, B. (1995). Limiting life-prolonging medical treatment: A comparative analysis of the President's Commission and the New York State Task Force. In R. Bulger, E. Bobby, & H. Fineberg (Eds.), *Society's choices: Social and ethical decision making in biomedicine* (pp. 307–374). Washington, DC: National Academy Press.

Bulger, R., Bobby, E., & Fineberg, H. (Eds.). (1995). *Society's choices: Social and ethical decision making in biomedicine.* Washington, DC: National Academy Press.

Butkus, M., & McCarthy, C. (2002). Principle and praxis: Harmonizing theoretical and clinical ethics. *American Journal of Bioethics, 2*(4). Retrieved October 31, 2005, from http://taylorandfrancis.metapress. com/link.asp?id=d0mtx3fru1t7d2xd

Butler, J., & Walter, J. (Eds.). (1991). *Transforming the curriculum: Ethnic studies and women's studies.* Albany, NY: State University of New York.

Byerly, R., Jr., & Pielke, R., Jr. (1995). The changing ecology of United States science. *Science, 269,* 1531–1532.

Callahan, D. (1998). Bioethics. In Reich, W. (Ed.), *Encyclopedia of bioethics* (pp. 247–256). New York: Macmillan Publishing Company.

Callahan, D. (1999). The Hastings Center and the early years of bioethics. *Kennedy Institute of Ethics Journal, 9*(1), 53–71.

Campbell, A. (2002). A lawyer's perspective on graduate studies in bioethics. *American Journal of Bioethics, 2*(4). Retrieved October 31, 2005, from http://taylorandfrancis.metapress.com/link.asp?id=mc959g5bvdpvrf5k

Campbell, E., Weissman, J., Clarridge, B., Yucel, R., Causino, N., & Blumenthal, D. (2003). Characteristics of medical school faculty members serving on institutional review boards: Results of a national survey. *Academic Medicine, 78*(8), 831–836.

Caplan, A. (1991, March/April). Bioethics on trial. *Hastings Center Report,* 19–20.

Chaitin, E. (2002). Earl Weaver was right: It's what you learn after you think you know it all that counts. *American Journal of Bioethics, 2*(4). Retrieved October 31, 2005, from http://taylorandfrancis.metapress. com/link.asp?id=2nhfh9c2ly1clypn

Chambliss, D. (1993). Is bioethics relevant? *Contemporary Sociology, 22*(5), 649–652.

Charo, R. (1995). La pénible valse hésitation: Fetal tissue research review and the use of bioethics commissions in France and the United States. In R. Bulger, E. Bobby, & H. Fineberg (Eds.), *Society's choices: Social and ethical decision making in biomedicine* (pp. 447–502). Washington, DC: National Academy Press.

Chubin, D., & Hackett, E. (1990). *Peerless science: Peer review and U.S. science policy*. Albany, NY: State University of New York Press.

Churchill, L. (1999). Are we professionals? A critical look at the social role of bioethicists. *Journal of the American Academy of Arts and Sciences, 128*(4), 253–274.

Clark, B. (2000). Developing a career in the study of higher education. In J. C. Smart (Ed.), *Higher education: Handbook of theory and research* (Vol. XV, pp. 1–38). Bronx, NY: Agathon Press.

Cleveland, H. (1964). Inquiry into presidential inquirers. In D. Johnson & J. Walker, (Eds.), *The dynamics of the American presidency* (pp. 291–294). New York: Wiley.

Collins, R. (1989). Toward a theory of intellectual change: The social causes of philosophies. *Science, Technology, & Human Values, 14*(2), 107–139.

Cooter, R. (1995). The resistible rise of medical ethics. *Social History of Medicine, 8*(2), 257–270.

Croissant, J. (2000). Critical legal theory and critical science studies. *Cultural Dynamics, 12*(2), 223–236.

Dembner, A. (2002, July/August). Lawsuits target medical research. *IRB: Ethics & Human Research*, 14–15.

Deverell, J. (1999, September 4). U of T appeals to Ottawa to help generic drug firms. *Toronto Star*. Retrieved April 21, 2008, from http://pqasb. pqarchiver.com/thestar/access/427902531.html?dids=427902531:427 902531&FMT=ABS&FMTS=ABS:FT&date=Sep+4%2C+1999& author=John+Deverell&pub=Toronto+Star&edition=&startpage=1& desc=U+of+T+appeals+to+Ottawa+to+help+generic+drug+firms

DeVries, R. (1995). Toward a sociology of bioethics. *Qualitative Sociology, 18*(1), 119–128.

DeVries, R., & Subedi, J. (1998). *Bioethics and society: Constructing the ethical enterprise*. Upper Saddle River, NJ: Prentice Hall.

Dickstein, E., Erlen, J., & Erlen, J. A. (1991). Ethical principles contained in currently professed medical oaths. *Academic Medicine, 66*(10), 622–624.

DiMaggio, P., & Powell, W. (1983). The iron cage revisited: Institutional isomorphism and collective rationality in organizational fields. *American Sociological Review, 48,* 147–160.

Dimond, K., & Alvarez, J. (2004). Investigator-initiated research: Proceed with caution. *Medical Research Law & Policy Report, 3*(14), 574–576.

EAB (Ethics Advisory Board, Department of Health Education and Welfare). (1979/1988). *Research on in vitro fertilization* (*Federal Register, 118,* 35033–35058). Reproduced in part in A. Jonsen, R. Veatch, & L. Walters (Eds.), *Source book in bioethics: A documentary history* (pp. 88–102). Washington, DC: Georgetown University Press.

Eckenwiler, L. (2001). Moral reasoning and the review of research involving human subjects. *Kennedy Institute of Ethics Journal, 11*(1), 37–69.

Egan, E. (2002). Ethics training in graduate medical education. *American Journal of Bioethics, 2*(4), 26–27.

Eisen, A., & Berry, R. (2002). The absent professor: Why we don't teach research ethics and what to do about it. *American Journal of Bioethics, 2*(4), 38–49.

Eiseman, E. (2003). *The National Bioethics Advisory Commission: Contributing to public policy.* A report conducted by RAND for the National Bioethics Advisory Commission. Santa Monica, CA: RAND Corporation.

Eisner, R. (1991). Institutions hustle to meet NIH ethics training mandate. *The Scientist, 6*(21), 1–25.

Elliott, C. (2001). Pharma buys a conscience. *American Prospect, 12*(17), 16–20.

Emanuel, E. J., Wood, A., Fleischman, A., Bowen, A., Getz, K. A., Grady, C., et al. (2004). Oversight of human participants research: Identifying problems to evaluate reform proposals. *Annals of Internal Medicine, 141*(4), 282–291.

Epstein, S. (1996). *Impure science: AIDS, activism, and the politics of knowledge.* Berkeley, CA: University of California Press.

Ettorre, E. (1999). Experts as 'storytellers' in reproductive genetics: Exploring key issues. *Sociology of Health & Illness, 21*(5), 539–559.

Evans, J. (1998). *Playing God? Human genetic engineering and the rationalization of bioethics 1959–1995.* Unpublished doctoral dissertation, Princeton University, NJ.

Evans, J. (2002). *Playing God? Human genetic engineering and the rationalization of public bioethical debate.* Chicago: University of Chicago Press.

Executive Order 12975. (1995, October 3). Protection of human research subjects and creation of National Bioethics Advisory Commission. *Federal Register, 60*(193), 52063–52065.

Executive Order 13237. (2001, November 28). Creation of the President's Council on Bioethics. *Federal Register, 66*(231), 59851.

Fairfield, H. (2007, September 12). Master's degrees abound as universities and students see a windfall. *New York Times.* Retrieved April 21, 2008, from http://www.nytimes.com/2007/09/12/education/12masters.html

Fallows, J. (1999, June 7). Profile: The political scientist. *The New Yorker*, p. 66.

Federman, D., Hanna, K., & Rodriguez, L., Eds. (2003). *Responsible research: A systems approach to protecting research participants.* Washington, DC: Institute of Medicine, National Academy Press.

FCEPRI (Florida Council for Educational Policy, Research and Improvement). (2003). *Public post-secondary centers and institutes.* Retrieved October 30, 2005, from http://www.cepri.state.fl.us/pdf/CI%20studyrevised.pdf

Fletcher, J. (1997). Bioethics in a legal forum: Confessions of an "expert" witness. *The Journal of Medicine and Philosophy, 22,* 297–324.

Fletcher, J. (2001). The stem cell debate in historical context. In S. Holland, K. Lebacqz, & L. Zoloth (Eds.), *The human embryonic stem cell debate: Science, ethics, and public policy* (pp. 27–34). Cambridge, MA: MIT Press.

Flitner, D. (1986). *The politics of presidential commissions.* Dobbs Ferry, NY: Transnational Publishing.

Flynn, P. (1991a). The disciplinary emergence of bioethics and bioethics committees: Moral ordering and its legitimation. *Sociological Focus, 24*(2), 145–56.

Flynn, P. (1991b). *Moral ordering and the social construction of bioethics.* San Francisco: University of California.

Fox, D. (1985). Who are we: The political origins of the medical humanities. *Theoretical Medicine and Bioethics, 6,* 327–341.

Fox, R. (1974). Ethical and existential developments in contemporaneous American medicine: Their implications for culture and society. *Milbank Quarterly, 52,* 445–483.

Fox, R. (1976). Advanced medical technology—Social and ethical implications. *Annual Review of Sociology, 2,* 231–268.

Fox, R., & Swazey, J. (1974). *The courage to fail: A social view of organ transplantation.* Chicago: Chicago University Press.

Fox, R., & Swazey, J. (1984). Medical morality is not bioethics—Medical ethics in China and the United States. *Perspectives in Biology and Medicine, 27*(3), 336–360.

Frader, J. (1992). Political and interpersonal aspects of ethics consultation. *Theoretical Medicine, 13*, 31–44.

Freidson, E. (1970). *Profession of medicine.* Chicago: University of Chicago Press.

Frohock, F. M. (1992). *Healing powers: Alternative medicine, spiritual communities, and the state.* Chicago: University of Chicago Press.

Fuchs, S. (1994). *The professional quest for truth: A social theory of scientific knowledge.* Albany, NY: State University of New York Press.

Gacki-Smith, J., & Gordon, E. (2005). Residents' access to ethics consultations: Knowledge, use, and perceptions. *Academic Medicine, 80*(2), 168–175.

Geertz, C. (1973). Thick description: Toward an interpretive theory of culture. In *The interpretation of cultures: Selected essays* (pp. 3–30). New York: Basic Books.

Geiger, R. (1990). Organizational research units—Their role in the development of university research. *Journal of Higher Education, 61*(1), 1–19.

Gelsinger v. University of Pennsylvania (Pa. C., 001885, complaint filed September 18, 2000). Retrieved October 26, 2005, from http://www.sskrplaw.com/links/healthcare2.html

Gieryn, T. (1983). Boundary work and the demarcation of science from non-science: Strains and interests in professional ideologies of scientists. *American Sociological Review, 48*, 781–795.

Gieryn, T. (1995). Boundaries of science. In S. Jasanoff, G. Markle, J. Petersen, & T. Pinch (Eds.), *Handbook of science and technology studies* (pp. 393–443). Thousand Oaks, CA: Sage Publications.

Gieryn, T. (1999). *Cultural boundaries of science: Credibility on the line.* Chicago: University of Chicago Press.

Goodman, K. (1999). Philosophy as news: Bioethics, journalism, and public policy. *Journal of Medicine and Philosophy, 24*(2), 181–200.

Gordon, E. (2002). The ASBH membership survey: Preliminary findings. *ASBH Exchange, 5*(3), 1, 6–7.

Gose, B. (2000, September 29). Penn doctors, ethicist named in suit over gene-therapy death. *Chronicle of Higher Education, 47*(2), A34.

Gottweis, H. (1998). *Governing molecules: The discursive politics of genetic engineering in Europe and the United States.* Cambridge, MA: The MIT Press.

Graham, H. (1985). The ambiguous legacy of American presidential commissions. *The Public Historian, 7*(2), 5–25.

Graumann, S. (2000). Bioethics or biopolitics? On the relationship between academic and public discussion of the "selection" and "manipulation" of human life. *Biomedical Ethics, 5*(3), 120–124.

Gray, B. (1995). Bioethics commissions: What can we learn from past successes and failures? In R. Bulger, E. Bobby, & H. Fineberg (Eds.), *Society's choices: Social and ethical decision making in biomedicine* (pp. 261–306). Washington, DC: National Academy Press.

Greenwall Foundation. (2005). *Bioethics program guidelines.* Retrieved October 10, 2005, from http://www.greenwall.org/exguide.html

Gregg v. Kane, 1997 WL 570909 (E.D. Pa. 1997).

Grimes v. Kennedy Krieger Inst., Inc., 782 A. 2d 807 (Md. 2001).

Gurin, J. (2002). A bull market for biomedical ethics. *American Journal of Bioethics, 2*(4), 35.

Guston, D. (2000). *Between politics and science.* New York: Cambridge University Press.

Guston, D., & Keniston, K. (1994). *The fragile contract: University science and the federal government.* Cambridge, MA: The MIT Press.

Hackett, E. (2001). Organizational perspectives on university-industry research relations. In J. Croissant & S. Restive (Eds.), *Degrees of compromise: Industrial interests and academic values* (pp. 1–21). Albany, NY: State University of New York.

Hafemeister, T., & Robinson, D. (1994). The views of the judiciary regarding life-sustaining medical treatment decisions. *Law & Psychology Review, 18*, 189–246.

Halfon, S. (1998). Collecting, testing, and convincing: Forensic DNA experts in the courts. *Social Studies of Science, 28*(5), 801–828.

Halikas v. University of Minnesota, 856 F. Supp. 1331 (D. Minn. 1994).

Hanson, M. (1999). Book review: Bioethics and Society. Journal of Value Inquiry, 33, 423–428.

Harrison, C. (2002). A Canadian perspective. *American Journal of Bioethics, 2*(4), 18–20.

Hastings Center. (1999, March/April). Symposium: Human Primordial Stem Cells. *Hastings Center Report 29*(2), 30–48.

Heilicser, B., Meltzer, D., & Siegler, M. (2000). The effect of clinical medical ethics consultation on healthcare costs. *Journal of Clinical Ethics, 11*, 31–38.

Heydebrand, W. (1989). New organizational forms. *Work and Occupations, 16*(3), 323–357.

HFTTR Panel (Human Fetal Tissue Transplantation Research Panel). (1988). *Report of the Human Fetal Tissue Transplantation Research Panel*. Bethesda, MD: National Institutes of Health.

Hoffman, D. (1991). Does legislating hospital ethics committees make a difference? A study of hospital ethics committees in Maryland, the District of Columbia, and Virginia. *Law, Medicine & Health Care, 19*(1-2), 105–19.

Hoffman, S., & Berg, J. (2005). *The suitability of IRB liability* (Case Research Paper Series in Legal Studies, Working paper 05-4). Case Western Reserve University, Cleveland, OH. Retrieved December, 20, 2007, from http://ssrn.com/abstract=671004

Hoke, F., & Kreeger, K. (1994). Arthur Caplan discusses issues facing the growing field of bioethics. *The Scientist, 8*(20), 12.

Hopkins, P. (1998). Bad copies: How popular media represent cloning as an ethical problem. *Hastings Center Report, 28*(2), 6–13.

Hsu, M. (1999, July). Banning human cloning: An acceptable limit on scientific inquiry or an unconstitutional restriction of symbolic speech? *Georgetown Law Journal.* Retrieved October 12, 2005, from http://www.findarticles.com/p/articles/mi_qa3805/is_199907/ai_n8876700

Icenogle, D. (2003). IRBs, conflict and liability: Will we see IRBs in court? Or is it when? *Clinical Medicine & Research, 1*(1), 63–68.

In re Quinlan, 70 NJ 10, 355 A.2d 647 (1976), *cert. denied*, 429 US 922 (1976).

Iserson, K., & Stocking, C. (1993a). Prevalences of ethics, socioeconomics, and legal education requirements in residency training. *Academic Medicine, 68*(1), 89–90.

Iserson, K., & Stocking, C. (1993b). Requirements for ethics, socioeconomic, and legal education in postgraduate medical programs. *Journal of Clinical Ethics, 4*(3), 225–229.

Isinger, M. (2002). The state of graduate education: One student's view. *American Journal of Bioethics, 2*(4). Retrieved April 21, 2008, from http://muse.jhu.edu/journals/american_journal_of_bioethics/v002/2.4isinger.pdf

Jansson, R. (2003). Research liability for negligence in human subject research: Informed consent and researcher malpractice actions. *Washington Law Review, 78*(1), 229–263.

Jasanoff, S. (1990). *The fifth branch: Science advisers as policymakers.* Cambridge, MA: Harvard University Press.

Jasanoff, S. (1996). Beyond epistemology: Relativism and engagement in the politics of science. *Social Studies of Science, 26*, 393–418.

Jasanoff, S. (1998). The eye of everyman: Witnessing DNA in the Simpson trial. *Social Studies of Science, 28*(5), 713–740.

JCAHO (Joint Commission on Accreditation for Healthcare Organizations). (1992). *Accreditation manual for hospitals: Standards* (Vol. I). Oakbrook, IL: Author.

JHU (Johns Hopkins University). (2001, July 19). "Hopkins responds to OHRP suspension of research." Press release, Johns Hopkins Medicine News & Information Services. Retrieved April 17, 2008, from http://www.hopkinsmedicine.org/press/2001/JULY/010719.htm

Jones, J. (1993). *Bad blood: The Tuskegee syphilis experiment* (Expanded ed.). New York: Free Press.

Jonsen, A. (1998a). *The birth of bioethics*. New York: Oxford University Press.

Jonsen, A. (1998b). The ethics of research with human subjects: A short history. In A. Jonsen, L. Walters, & R. Veatch (Eds.), *Source book in bioethics: A documentary history* (pp. 5–10). Washington, DC: Georgetown University Press.

Kay, L. (1993). *The molecular vision of life: Caltech, the Rockefeller Foundation, and the rise of the new biology*. New York: Oxford University Press.

Kelly, S. E. (1994). *Moral boundaries of medical research: A sociological analysis of human fetal tissue transplantation*. Unpublished doctoral dissertation, University of California, San Francisco.

Kelly, S. E. (2003). Public bioethics and publics: Consensus, boundaries, and participation in biomedical science policy. *Science, Technology, & Human Values, 28*(3), 339–364.

Kevles, D. (1995). *The physicists: The history of a scientific community in modern America*. Cambridge, MA: The MIT Press.

Knorr-Cetina, K. (1981). *The manufacture of knowledge: An essay on the constructivist and contextual nature of science*. Oxford, U.K.: Pergamon Press.

Kreeger, K. (1994). Burgeoning crop of bioethics programs and courses reflects the deepening of scientists' moral concerns. *The Scientist* (8), 1.

Kuczewski, M., & Parsi, K. (2002). The virtual graduate program in bioethics: The mission, the students, and the hazards. *American Journal of Bioethics, 2*(4), 13–17.

Kus v. Sherman Hospital, 644 N.E.2d 1214 (Ill. App. 1995).

Lanouette, W. (1981, December 12). Nuclear committee plays it straight-and draws criticism from all quarters. *National Journal*, pp. 2203–2205.

Lantos, J. (2002). Minding our own store. *ASBH Exchange, 5*(3), 2, 13.

Lanza, R., et al. (1999). Science over politics [Letter to the editor]. *Science, 283*, 1849–1850.

LaPuma, J., Stocking, C., Silverstein, M., DiMartini, A., & Siegler, M. (1988). An ethics consultation service in a teaching hospital: Utilization and evaluation. *Journal of the American Medical Association, 260*, 808–811.

Larson, M. (1977). *The rise of professionalism: A sociological analysis.* Berkeley, CA: University of California Press.

Larson, M. (1990). In the matter of experts and professionals, or how impossible it is to leave nothing unsaid. In R. Torstendahl & M. Burrage (Eds.), *The formation of professions: Knowledge, state, and strategy* (pp. 24–50). Newbury Park, CA: Sage Publications.

Larson, R., & Barnes-Moorhead, S. (2001). *How centers work: Building and sustaining academic nonprofit centers.* A publication of the W.K. Kellogg Foundation. Retrieved October 30, 2005, from http://www.wkkf.org/Pubs/PhilVol/BuildingBridges/Pub3414.pdf

Latour, B., & Woolgar, S. (1986). *Laboratory life: The construction of scientific facts.* Princeton, NJ: Princeton University Press.

Leinhos, M. (2002a, November 7–9). *Is bioethics "just politics"?* Paper presented at the Society for the Social Studies of Science 26th annual meeting, Milwaukee, WI.

Leinhos, M. (2002b, November). *The function and impact of federal bioethics bodies.* Invited paper, Research Symposium With the Next

Generation of Leaders in Science and Technology Policy, Washington, DC. Retrieved from October 31, 2005, from http://www.cspo.org/nextgen/Leinhos.PDF

Leslie, S. (1993). *The cold war and American science: The military-industrial-academic complex at MIT and Stanford.* New York: Columbia University Press.

Light, D., & McGee, G. (1998). On the social embeddedness of bioethics. In R. DeVries & J. Subedi (Eds.), *Bioethics and society: Constructing the ethical enterprise* (pp. 1–15). Upper Saddle River, NJ: Prentice Hall.

Lipsky, M., & Olson, D. (1976). *Commission politics: The processing of racial crisis in America.* New Brunswick, NJ: Transaction Books.

Lock, M. (1996). Death in technological time: Locating the end of meaningful life. *Medical Anthropology Quarterly, 10*(4), 575–600.

Lock, M. (2001). *Twice dead: Organ transplants and the reinvention of death.* Berkeley, CA: University of California Press.

Lockwood Tooher, N. (2005, September 2). Clinical trial lawsuits are on the rise. *Lawyers Weekly, Inc.* Retrieved October 22, 2005, from http://www.sskrplaw.com/publications/050902.html

Lynch, M., & Woolgar, S. (Eds.). (1990). *Representation in scientific practice.* Cambridge, MA: The MIT Press.

Magnus, D. (2002). The meaning of graduate education for bioethics. *American Journal of Bioethics, 2*(4), 10–12.

Maloney, D. (2003). In court: Court says the institutional review board (IRB) followed correct procedures. *Human Research Report, 18*(5), 8.

Marcus, G. (1998). *Ethnography through thick and thin.* Princeton, NJ: Princeton University Press.

Marker, R. (2001, January/February). Dying for the cause: Foundation funding for the 'right-to-die' movement. *Philanthropy Magazine*, 26–29.

Marshall, E. (1998, November 11). A versatile cell line raises scientific hopes, legal questions. *Science, 282*, 1014–1015.

Marshall, E. (1999, July 23). Ethicists back stem cell research, White House treads cautiously. *Science, 285*, 502.

Martin, E. (1994). *Flexible bodies: Tracking immunity in American culture from the days of polio to the age of AIDS*. Boston: Beacon Press.

McAllen, P., & Delgado, R. (1984, February). Moral experts in the courtroom. *Hastings Center Report*, 27–34.

McCarrick, P. (1993). Scope note 33: Bioethics consultation. *Kennedy Institute of Ethics Journal, 3*(4), 433–451.

McCarthy, E. (1996). *Knowledge as culture: The new sociology of knowledge*. New York: Routledge.

McClung, J., Kamer, R., DeLuca, M., & Barber, H. (1996). Evaluation of a medical ethics consultation service: Opinions of patients and healthcare providers. *American Journal of Medicine, 100*, 456–460.

McCruden, P. (2002). Why an online graduate bioethics program: One student's experience. *American Journal of Bioethics, 2*(4), 25.

McGee, G., Spanogle, J., Caplan, A., Penny, D., & Asch, D. (2002). Success and failures of hospital ethics committees: A national survey of ethics committee chairs. *Cambridge Quarterly of Healthcare Ethics, 11*, 87–93.

McGuire, A. (2002). Clearing the mist. *American Journal of Bioethics, 2*(4). Retrieved October 31, 2005, from http://taylorandfrancis.metapress.com/link.asp?id=g81yavjvc5g4ye31

Meehan, M. (1996, November 3). Looking more like America? *Our Sunday Visitor*. Retrieved December 4, 2001, from http://www.catholic.net/RCC/Periodicals/OSV/96nov3.html

Meilaender, G. (1995). *Body, soul, and bioethics*. Notre Dame, IN: University of Notre Dame Press.

Mello, M., Studdert, D., & Brennan, T. (2003). The rise of litigation in human subjects research. *Annals of Internal Medicine, 139*(1), 40–45.

Meyer, M., Genel, M., Altman, R., Williams, M., & Allen J. (1998). Clinical research: Assessing the future in a changing environment: Summary report of conference sponsored by the American Medical Association Council on Scientific Affairs, Washington, DC, March, 1996. *American Journal of Medicine, 104*(3), 264–271.

Moore, K. (1996). Organizing integrity: American science and the creation of public interest organizations, 1955–1975. *American Journal of Sociology, 101*(6), 1592–1627.

Moreno, J. (1995). *Deciding together: Bioethics and moral consensus.* New York: Oxford University Press.

Moreno, J. (2004). Letter from the president: Beyond the annual meeting. *ASBH Exchange, 7*(2), 2.

Moreno, J., & Hurt, V. (1998). How the Atomic Energy Commission discovered "informed consent." In R. DeVries & J. Subedi (Eds.), *Bioethics and society: Constructing the ethical enterprise* (pp. 78–93). Upper Saddle River, NJ: Prentice Hall.

Morreim, E. H. (2004, fall) Litigation in clinical research: malpractice doctrines versus research realities. *Journal of Law, Medicine & Ethics, 32*(3), 474–484.

Muller, J. (1994). Anthropology, bioethics, and medicine: A provocative trilogy. *Medical Anthropology Quarterly, 8*(4), 448–467.

Mullins, N. (1972). The development of a scientific specialty: The Phage Group and the origins of molecular biology. *Minerva, 19*, 52–82.

Nathanson, J. (2002). Medical ethics and the moral practice of medicine. *American Journal of Bioethics, 2*(4). Retrieved October 31, 2005, from http://taylorandfrancis.metapress.com/link.asp?id=74v7a5jg174wbtml

National Commission (National Commission for the Protection of Human Subjects of Biomedical and Behavioral Research). (1979). *The Belmont report: Ethical principles and guidelines for the protection of human subjects of research.* Washington, DC: U.S. Government Printing Office.

National Public Radio. (1996). *David Baron reports on a new federal bioethics commission*. A report featured on the program "Morning Edition." Retrieved November 1, 2001, from http://www.npr.org/templates/story/story.php?storyId=1030461

National Research Council (NRC). (1995). *Research-doctorate programs in the United States: Continuity and change*. Washington DC: National Academy Press. Retrieved March 12, 2006, from http://newton.nap.edu/html/researchdoc/

NBAC (National Bioethics Advisory Commission). (1999a). *Ethical issues in human stem cell research* (Report and recommendations). Rockville, MD.

NBAC (National Bioethics Advisory Commission). (1999b). *Transcript of NBAC meeting of January, 19, 1999*. Bethesda, MD: Author.

NBAC (National Bioethics Advisory Commission). (1999c). *Transcript of NBAC meeting of February 2, 1999*. Bethesda, MD: Author.

NBAC (National Bioethics Advisory Commission). (1999d). *Transcript of NBAC meeting of February 3, 1999*. Bethesda, MD: Author.

NBAC (National Bioethics Advisory Commission). (1999e). *Transcript of NBAC meeting of April 16, 1999*. Bethesda, MD: Author.

NBAC (National Bioethics Advisory Commission). (1999f). *Transcript of NBAC meeting of June 28, 1999*. Bethesda, MD: Author.

NBAC (National Bioethics Advisory Commission). (2000). *National Bioethics Advisory Commission 1998–1999 biennial report*. Bethesda, MD: Author.

NCSL (National Conference of State Legislatures). (2008). *State human cloning laws. A listing of state cloning laws, updated January, 2008*. Retrieved April 21, 2008, from http://www.ncsl.org/programs/health/genetics/rt-shcl.htm

Nelkin, D. (1995). *Selling science: How the press covers science and technology*. New York: WH Freeman.

OIG (Office of the Inspector General, U.S. Department of Health and Human Services). (1998). *Institutional review boards: A time for reform* (DHHS Publication No. OEI-01-97-00193). Washington, DC: U.S. Government Printing Office.

Orr, R., & Moon, E. (1993). Effectiveness of an ethics consultation service. *Journal of Family Practice, 36,* 49–53.

OTA (Office of Technology Assessment, U.S. Congress). (1993). *Biomedical ethics in U.S. public policy—Background paper* (DHHS Publication No. OTA-BP-BBS-105). Washington, DC: U.S. Government Printing Office.

President's Commission (President's Commission for the Study of Ethical Problems in Medicine and Biomedical and Behavioral Research). (1983). *Deciding to forego life-sustaining treatment: A report on the ethical medical, and legal issues in treatment decisions* (DHHS Publication No. O-402-884). Washington, DC: Government Printing Office.

Page, C. (1996). A passion for the mundane, and our medical education and threatened teaching hospitals. *American Journal of Surgery, 172,* 398–404.

Paris, J. (1984). An ethicist takes the stand. *Hastings Center Report* (February, 1984), 32–33.

Peirce, A. (2004, October 14). Some considerations about decisions and decision-makers in hospital ethics committees. *Online Journal of Health Ethics, 1*(1). Retrieved April 10, 2006, from http://ethicsjournal. umc.edu/ojs/viewarticle.php?id=15

Perlstadt, H. (2004). The researcher's bill of rights. *Medical Humanities Report, 25*(4). Retrieved October, 25, 2005, from http://bioethics.msu. edu/mhr/04w/perlstadt.html

Pfeffer, J., & Salancik, G. (1978). *The external control of organizations.* New York: Harper and Row.

Poland, S. (1997). Scope note 33: Landmark legal cases in bioethics. *Kennedy Institute of Ethics Journal, 7*(2), 191–209.

Pope, T. (2002). My bioethics education at Georgetown. *American Journal of Bioethics, 2*(4), 36.

Potter, V. (1970). Bioethics, the science of survival. *Perspectives in Biology and Medicine, 14*, 127–153.

Rado, L. (1987). Cultural elites and the institutionalization of ideas. *Sociological Forum, 2*(1), 42–66.

Rapp, R. (1998). Refusing prenatal diagnosis: The meanings of bioscience in a multicultural world. *Science, Technology, & Human Values, 23*(1), 45–70.

Rapp, R. (1999). *Testing women, testing the fetus: The social impact of amniocentesis in America.* New York: Routledge.

Rawls, J. (1987). The idea of overlapping consensus. *Oxford Journal of Legal Studies, 7*(1), 125.

Resnik, D. (2004). Liability for institutional review boards: From regulation to litigation. *Journal of Legal Medicine, 25*, 131–184.

Robertson v. McGee, 2002 WL 535045 (N.D. Okla. 2002).

Rothman, D. (1991). *Strangers at the bedside: A history of how law and bioethics transformed medical decision making.* New York: Basic Books.

Russo, E. (1999). 'Bioethicists' proliferate despite undefined career track. *The Scientist, 13*(8), 16.

Rycroft, R. (1991). Environmentalism and science: Politics and the pursuit of knowledge. *Knowledge: Creation, Diffusion, Utilization, 13*(2), 150–169.

Scheirton, L., & Kissell, J. (2001). The leverage of the law: The increasing influence of law on healthcare ethics committees. *HEC Forum, 13*(1), 1–12.

Schmitt, G. J. (1989, spring). Why commissions don't work. *The National Interest*, pp. 58–66.

Schneider, A. (1999, May 21). Master's degrees, once scorned, attract students and generate revenue. *Chronicle of Higher Education*, pp. A12–13.

Schneiderman, L., Gilmer, T., Teetzel, H., Dugan, D., Blustein, J., Cranford, R., et al. (2003). Effect of ethics consultations on nonbeneficial life-sustaining treatments in the intensive care setting: A randomized controlled trial. *Journal of the American Medical Association, 290*, 1166–1172.

Schonfeld, T. (2002). Balancing bioethics. *American Journal of Bioethics, 2*(4), 32–33.

Sepinwall, A. (2002). The merits of a general education in bioethics. *American Journal of Bioethics, 2*(4), 31.

Shalit, R. (1997). When we were philosopher kings. *The New Republic, 216*(17), 24–28.

Shamblott, M., Axelman, J., Wang, S., Bugg, E. M., Littlefield, J., Donovan, P. J., et al. (1998). Derivation of pluripotent stem cells from cultured human primordial germ cells. *Proceedings of the National Academy of Sciences of the United States of America, 95*, 13726–13731.

Shaul, R. Z., Birenbaum, S., & Evans, M. (2005, June 13). Legal liabilities in research: Early lessons from North America. *BMC Medical Ethics, 6*, E4. Retrieved April 21, 2008, from http://www.biomedcentral.com/content/pdf/1472-6939-6-4.pdf

Shivas, T. (2002). A dash of this and a pinch of that: The role of interdisciplinary opportunities in graduate education. *American Journal of Bioethics, 2*(4), 24.

Simonson, P. (2002). Bioethics and the rituals of the media. *Hastings Center Report, 32*(1), 32–39.

Sismondo, S. (1996). *Science without myth: On constructions, reality, and social knowledge*. Albany, NY: State University of New York Press.

Sisti, D. (2002). There he is…Master of bioethics. *American Journal of Bioethics, 2*(4), 28–29.

Slaughter, S. (1990). *The higher learning and high technology: Dynamics of higher education policy formation.* Albany, NY: State University of New York Press.

Slaughter, S. (1991). The "official" ideology of higher education: Ironies and inconsistencies. In W. Tierney (Ed.), *Culture and ideology in higher education: advancing a critical agenda* (pp. 59–85). New York: Praeger Publishers.

Slaughter, S. (1997). Class, race and gender and the construction of postsecondary curricula in the United States. *Journal of Curriculum Studies, 29*(1), 1–30.

Slaughter, S., & Leslie, L. (1997). *Academic capitalism.* Baltimore: Johns Hopkins University Press.

Slaughter, S., & Rhoades, G. (1996). The emergence of a competitiveness research and development policy coalition and the commercialization of academic science and technology. *Science, Technology, & Human Values, 21*(3), 3303–3339.

Slaughter, S., & Rhoades, G. (2004). *Academic capitalism and the new economy: Markets, state, and higher education.* Baltimore: Johns Hopkins University Press.

Small, M. (1999). Departmental conditions and the emergence of new disciplines: Two cases in the legitimation of African-American studies. *Theory and Society 28*, 659–707.

Smith, B. (1990). *American science policy since World War II.* Washington, DC: Brookings Institution.

Solomon, R. (2002). The value of bioethics education. *American Journal of Bioethics, 2*(4). Retrieved October 21, 2005, from http://taylorandfrancis. metapress.com/link.asp?id=rxkm3hjfjnthdnl3

Solomon, S., & Hackett, E. (1996). Setting boundaries between science and law: Lessons from *Daubert v. Merrell Dow Pharmaceuticals, Inc. Science, Technology, & Human Values, 21*(2), 131–156.

Sontag, D. (2002). Are clinical ethics consultants in danger? An analysis of the potential legal liability of individual clinical ethicists. *University of Pennsylvania Law Review, 151*(2), 667–705.

Spielman, B. (2001, fall). Has faith in health care ethics consultants gone too far? Risks of an unregulated practice and a model act to contain them. *Marquette Law Review, 85*, 161–221.

Spielman, B., & Agich, G. (1999). The future of bioethics testimony: Guidelines for determining qualifications, reliability, and helpfulness. *San Diego Law Review, 36*, 1043–1075.

Stahler, G., & Tash, W. (1994). Centers and institutes in the research university: Issues, problems, and prospects. *Journal of Higher Education, 65*(5), 540–554.

Star, S., & Griesemer, J. (1989). Institutional ecology, translations, and boundary objects: Amateurs and professionals in Berkeley's Museum of Vertebrate Zoology, 1907–39. *Social Studies of Science, 19*, 387–420.

Stevens, M. (2000). *Bioethics in America: Origins and cultural politics.* Baltimore: Johns Hopkins University Press.

Stolberg, S. (2001, August 2). Bioethicists find themselves the ones being scrutinized. *New York Times*, pp. A1, A14.

Strauss, M. (2002). The place of philosophy. *American Journal of Bioethics, 2*(4). Retrieved October 31, 2005, from http://taylorandfrancis.metapress.com/link.asp?id=jwk0ypugyeeb11k3

Sulzner, G. (1974). The policy process and the uses of national government study commissions. In S. Bach & G. Sulzner (Eds.), *Perspectives on the presidency: A collection* (pp. 214–229). Lexington, MA: D.C. Heath.

Suziedelis, A. (2002). Topsy's midlife career in healthcare ethics. *American Journal of Bioethics, 2*(4). Retrieved October, 31, 2005, from http://taylorandfrancis.metapress.com/link.asp?id=yu9dhe1tap6hmy0w

Thomson, J. A., Itskovitz-Eldor, J., Shapiro, S. S., Waknitz, M. A., Swiergiel, J. J., Marshall, V. S., et al. (1998). Embryonic stem cell lines derived from human blastocysts. *Science, 282*, 1145–1147.

Tutchings, T. (1979). *Rhetoric and reality: Presidential commissions and the making of public policy.* Boulder, CO: Westview Press.

Veikher, E. (2002). An international student's perspective. *American Journal of Bioethics, 2*(4), 30.

Vogel, G. (2001). Can adult stem cells suffice? *Science, 292*, 1820–1822.

Voss, K. (2002). One field, many disciplines. *American Journal of Bioethics, 2*(4). Retrieved October, 31, 2005, from http://taylorandfrancis.metapress.com/link.asp?id=098l24pybtvpy9n4

Wade, N. (1998, November 12). Researchers claim embryonic cell mix of human and cow. *New York Times*, p. A1.

Wadman, M. (1997). Business booms for guides to biology's moral maze. *Nature, 389*, 658–659.

Washburn, J. (2005). *University, Inc.: The corporate corruption of American higher education.* New York: Basic Books.

Weisbard, A. (1987, July). The role of philosophers in the public policy process: A view from the President's Commission. *Ethics, 97*, 776–785.

Weiss, R., & Nelson, D. (2000, November 4). Penn settles gene therapy suit: University pays undisclosed sum to family of teen who died. *Washington Post*, p. A4.

Wetherill v. University of Chicago, 565 F. Supp. 1553, 1564 (N.D. Ill. 1983).

White v. Paulsen, 997 F. Supp. 1380, 1383 (E.D. Wash. 1998).

White, M. (2002). Why not medical humanities? *American Journal of Bioethics, 2*(4), 34.

Whitlock v. Duke University, 637 F. Supp. 1463 (M.D.N.C. 1986).

Winner, L. (2001). The gloves come off: Shattered alliances in science and technology studies. Reprinted in J. Croissant & S. Restivo (Eds.), *Degrees of compromise: Industrial interests and academic values* (pp. 241–251). Albany, NY: State University of New York Press.

Wolanin, T. (1975). *Presidential advisory commissions: Truman to Nixon.* Madison, WI: University of Wisconsin Press.

Wolpe, P. (1998). The triumph of autonomy in American bioethics: A sociological view. In R. DeVries & J. Subedi (Eds.), *Bioethics and society: Constructing the ethical enterprise* (pp. 38–59). Upper Saddle River, NJ: Prentice Hall.

Wolpe, P. (2000). From bedside to boardroom: Sociological shifts and bioethics. *HEC Forum, 12*(3), 191–201.

Wolpe, P., & McGee, G. (2001). "Expert bioethics" as professional discourse: The case of stem cells. In S. Holland, K. Lebacqz, & L. Zoloth (Eds.), *The human embryonic stem cell debate: Science, ethics, and public policy* (pp. 185–196). Cambridge, MA: The MIT Press.

Wright v. Fred Hutchinson Cancer Center, WL 32124953 (W.D. Wash. 2002).

Wright, S. (1994). *Molecular politics: Developing American and British regulatory policy for genetic engineering, 1972–1982.* Chicago: University of Chicago Press.

Zilberberg, J. (2002). Self-directed bioethics education. *American Journal of Bioethics, 2*(4). Retrieved October 31, 2005, from http://taylorandfrancis.metapress.com/link.asp?id=6rrvrufmdcbgamkq

Zoloth, L. (2001). Seeing the duties to all. *Hastings Center Report, 31*(2), 15–19.

Zoubul, C. (2002). Why study bioethics? Because it's interesting. *American Journal of Bioethics, 2*(4). Retrieved October 31, 2005, from http://taylorandfrancis.metapress.com/link.asp?id=xf5mnjq6xlyayw5a

Zussman, R. (1992). *Intensive care: Medical ethics and the medical profession.* Chicago: University of Chicago Press.

INDEX